SpringerBriefs in Molecular Science

Chemistry of Foods

Series editor

Salvatore Parisi, Al-Balqa Applied University, Al-Salt, Jordan

The series Springer Briefs in Molecular Science: Chemistry of Foods presents compact topical volumes in the area of food chemistry. The series has a clear focus on the chemistry and chemical aspects of foods, topics such as the physics or biology of foods are not part of its scope. The Briefs volumes in the series aim at presenting chemical background information or an introduction and clear-cut overview on the chemistry related to specific topics in this area. Typical topics thus include:

- Compound classes in foods—their chemistry and properties with respect to the foods (e.g. sugars, proteins, fats, minerals, ...)
- Contaminants and additives in foods—their chemistry and chemical transformations
- Chemical analysis and monitoring of foods
- Chemical transformations in foods, evolution and alterations of chemicals in foods, interactions between food and its packaging materials, chemical aspects of the food production processes
- Chemistry and the food industry—from safety protocols to modern food production

The treated subjects will particularly appeal to professionals and researchers concerned with food chemistry. Many volume topics address professionals and current problems in the food industry, but will also be interesting for readers generally concerned with the chemistry of foods. With the unique format and character of SpringerBriefs (50 to 125 pages), the volumes are compact and easily digestible. Briefs allow authors to present their ideas and readers to absorb them with minimal time investment. Briefs will be published as part of Springer's eBook collection, with millions of users worldwide. In addition, Briefs will be available for individual print and electronic purchase. Briefs are characterized by fast, global electronic dissemination, standard publishing contracts, easy-to-use manuscript preparation and formatting guidelines, and expedited production schedules.

Both solicited and unsolicited manuscripts focusing on food chemistry are considered for publication in this series. Submitted manuscripts will be reviewed and decided by the series editor, Dr. Salvatore Parisi.

To submit a proposal or request further information, please contact Dr. Sofia Costa, Publishing Editor, via sofia.costa@springer.com or Dr. Salvatore Parisi, Book Series Editor, via drparisi@inwind.it or drsalparisi5@gmail.com

More information about this series at http://www.springer.com/series/11853

Suresh D. Sharma · Arpan R. Bhagat ·
Salvatore Parisi

Raw Material Scarcity
and Overproduction
in the Food Industry

 Springer

Suresh D. Sharma
Department of Biochemistry
and Molecular Biology
Pennsylvania State University
University Park, State College, PA, USA

Arpan R. Bhagat
Saputo Dairy Foods USA, LLC
Dallas, TX, USA

Salvatore Parisi
Al-Balqa Applied University
Al-Salt, Jordan

ISSN 2191-5407   ISSN 2191-5415   (electronic)
SpringerBriefs in Molecular Science
ISSN 2199-689X   ISSN 2199-7209   (electronic)
Chemistry of Foods
ISBN 978-3-030-14650-4   ISBN 978-3-030-14651-1   (eBook)
https://doi.org/10.1007/978-3-030-14651-1

Library of Congress Control Number: 2019933841

© The Author(s), under exclusive license to Springer Nature Switzerland AG 2019
This work is subject to copyright. All rights are solely and exclusively licensed by the Publisher, whether the whole or part of the material is concerned, specifically the rights of translation, reprinting, reuse of illustrations, recitation, broadcasting, reproduction on microfilms or in any other physical way, and transmission or information storage and retrieval, electronic adaptation, computer software, or by similar or dissimilar methodology now known or hereafter developed.
The use of general descriptive names, registered names, trademarks, service marks, etc. in this publication does not imply, even in the absence of a specific statement, that such names are exempt from the relevant protective laws and regulations and therefore free for general use.
The publisher, the authors and the editors are safe to assume that the advice and information in this book are believed to be true and accurate at the date of publication. Neither the publisher nor the authors or the editors give a warranty, expressed or implied, with respect to the material contained herein or for any errors or omissions that may have been made. The publisher remains neutral with regard to jurisdictional claims in published maps and institutional affiliations.

This Springer imprint is published by the registered company Springer Nature Switzerland AG
The registered company address is: Gewerbestrasse 11, 6330 Cham, Switzerland

Contents

Chapter 1
Seasonal Variation and Biochemical Composition of Fishmeal

Abstract Nowadays, the food industry is forced to accept several challenges on a global scale in relation to the increased food demand for human consumption and correlated consequences. This chapter describes the observed variations of biochemical features of fishmeal in function of seasonal cycles. Main nutritional parameters may vary depending on the season and the geographical location; improper processing/storage treatments can worsen the product. For these reasons, the variability of fishmeal should be preliminarily examined on a seasonal basis. Fishmeal is considered as a protein-prevailing feeding material; the composition of fish feed should consider fishmeal ranging from 5 to 50%. In addition, a notable part of fish feed may be partially or totally replaced with vegetable proteins such as soy proteins. Moisture and lipid contents, the composition of fatty acids, and the presence of trace metals, antioxidants, and vitamins—correlated with 'freshness'—seem to depend on seasonal periods. Consequently, a strict routine control on fishmeal is critical, and the importance of affordable, easy-to-use, and rapid methods has to be evaluated when speaking of industrial plants instead of separated analytical laboratories.

Keywords Astaxanthin · Fishmeal · Lipid oxidation · Seafood · Soy protein · Unsaturated fatty acid · α-Tocopherol

Abbreviations

CO	Carbon monoxide
DDT	Dichlorodiphenyltrichloroethane
EFSA	European Food Safety Authority
H_2S	Hydrogen sulphide
NIRS	Near infrared reflectance spectroscopy
NO	Nitrogen monoxide
OI	Organochlorine insecticide
PCB	Polychlorinated biphenyl
PUFA	Polyunsaturated fatty acid
SO_2	Sulphur dioxide

© The Author(s), under exclusive license to Springer Nature Switzerland AG 2019
S. D. Sharma et al., *Raw Material Scarcity and Overproduction in the Food Industry*,
Chemistry of Foods, https://doi.org/10.1007/978-3-030-14651-1_1

1.1 Fishmeal Derived from Fish and Seafood in the Current Seafood Industry

One of the recurring and more important problems in the seafood industry is the variability of used fishmeal during the year. In fact, the observed variations of biochemical features appear to depend on peculiar seasonal cycles. Main nutritional parameters may vary depending on the season and the geographical location; improper processing/storage treatments can worsen the product. For these reasons, the variability of fishmeal should be preliminarily examined on a seasonal basis.

Generally, fishmeal is considered as a protein-prevailing feeding material, and the interested species—as fishmeal consumers—are normally carnivorous fish (eel, salmon, trout, carp, catfish, etc.). Roughly, the composition of fish feed should consider fishmeal ranging from 5 to 50%. Shrimp feed could be considered as a possible exception because animal protein amounts should not be lower than 30%, with a maximum amount of 50% (Dersjant-Li 2002; Drew et al. 2007; Elangovan and Shim 2000; Wang et al. 2006). In addition, a notable part of fish feed may be replaced with vegetable proteins such as soy proteins (minimum addition: 40%, maximum amount: 100%, or complete substitution). For these reasons, the amount of nitrogen-based molecules (proteins) has to be high, while oil quantities should be lower. With specific reference to fishmeal obtained from fish and seafood only, the problem concerns oil matters.

The quality and associated acceptability of fishmeal can depend on the following factors (Barlow 1994; Bragadóttir et al. 2004a; Hultin 1992; Ólafsson 1953; Opstvedt 1975; Pike et al. 1990; Romoser et al. 1968; Undeland 1997; Waissbluth et al. 1971):

(1) Freshness of the initial raw material(s)
(2) Processing of the initial raw material(s)
(3) Storage of intermediate products and the final fishmeal
(4) Biological and chemical variations of the initial raw material(s), including the reduced presence of natural antioxidants
(5) High amount of unsaturated fatty acids, with consequent high tendency to lipid oxidation, at normal and refrigerated storage conditions
(6) Partial demolition and decrease in nutritional quality of proteins (this phenomenon depends on lipid oxidation)
(7) Stability of the used fish.

Interestingly, the first three factors do not depend on the used fish, while all other points concern both the typology of used raw materials and seasonal features. From the chemical viewpoint, the following features should be considered and evaluated in relation to seasonal periods (Astrup and Halvorsen 1985; Opstvedt 1985; Waissbluth et al. 1971):

(1) Lipid amount
(2) Percentage content of fatty acids: total saturated, total monoene fatty acids, and polyunsaturated fatty acids (PUFA)

(3) Moisture content (the higher the water, the lower the resistance against oxidation)
(4) Iodine amount
(5) Quality of trace metals
(6) α-Tocopherol content
(7) Density of the fishmeal (probably, this factor is influenced by the amount of soluble substances in water).

With reference to processing and storage parameters, they can influence and modify the seasonal features of fishmeal. In particular, the following parameters should be considered (Waissbluth et al. 1971):

(a) Lipid variation
(b) Astaxanthin amount
(c) α-Tocopherol content
(d) Vitamin A
(e) Quantity of free fatty acids
(f) pH value
(g) Colorimetric appearance (or browning defects)
(h) Variation of emitted carbon monoxide (CO), hydrogen sulphide (H_2S), nitrogen monoxide (NO), and sulphur dioxide (SO_2).

Other factors could be taken into account. However, a good overview of the problem represented by fishmeal for food production purposes can be considered on these bases. The following sections concern the specific factors and the possible influence of seasonal variations on obtained and reported results.

1.2 Fishmeal and Variability. A Seasonal Viewpoint

The difference between fish species should be considered in terms of:

(a) Natural behaviour
(b) Position in the food web
(c) Chemical composition of fish tissues.

Interestingly, some studies have investigated the reliability of different fish species as fishmeal depending especially on the amount of oils (the higher the fat content, the lower the residual amount of protein and minerals) and related compound linked to the lipid phase, vitamin A in particular. Seasonal differences are not extremely evident: on the other side, the fish species seem to be the prevailing factor when speaking of adaptability for fishmeal purposes. Consequently, it has been reported that certain Australian species—including Bluefin tuna, salmon, and sea mullet—can be used as valuable sources of oils and vitamin A, while other fish—e.g., deep-sea flathead, garfish, snapper shark, etc.—are more useful as fishmeal because there is no need of processing except gutting. The seasonal variation can be evident enough, but the first factor seems the classification of fish species based on the simple examination of oil content, protein, ash, and calculated kilocalories (Jowett and Davies 1938).

In particular, the content of lipids has to be evaluated seasonally. The Tester's fat factor seems one of the most promising methods in this ambit, because of the possible evaluation of oil contents with and indirect equation: 'fish weight/water weight'. In this way, the oil amount of certain fish species can be evaluated season by season, or month by month, demonstrating that oil quantities may vary with notable deviation standards during all the collection year (McBride et al. 1959).

The problem of storage should be considered in connection with seasonal variations (Boran et al. 2008). In general, the problem seems more linked to the composition and related modifications of fish oil during time, but the seasonal variation of these features depends on:

(a) The natural variability of chemical composition during the year, according to above discussed factors correlated with fish species and their position in the food web, and

(b) The influence of storage conditions.

The natural variability of chemical composition does not concern only water, protein, fat, and ash amount in fish. It has been reported that Turkish anchovies may exhibit notable differences, and the variation of fish lipids ($\leq 6.3\%$) during the year may correspond to the analogous variation of water ($\leq 8.1\%$) (Boran et al. 2008).

On the other side, the modification of fish composition and quality (including also sanitary acceptability) depends on the amount of natural antioxidants. Generally, these molecules such as α-tocopherol and astaxanthin (one of carotenoids naturally present in salmons and rainbow trouts) are more abundant in the summer and spring seasons, while a remarkable diminution may be observed in winter and autumn, when speaking of certain fish types such as capelin (Bragadóttir et al. 2004b; Buttle et al. 2001; Nickell and Bromage 1998; Tolasa et al. 2005). α-Tocopherol (Fig. 1.1) has been reported and studied recently when speaking of fishmeal supplementation in different ambits (Webb et al. 1973). In particular, this antioxidant is considered a good option with reference to fishmeal stabilisation during shipping, with other natural mixtures such as rosemary extract, ethoxyquin, and/or butylated hydroxytoluene (UNECE 2016). Moreover, the 'International Maritime Dangerous Goods Code' of the International Maritime Organization considers fishmeal a hazardous cargo (Aquaculture Working Group 2012). Therefore, mixed tocopherols may be used and added to fishmeal to prevent combustion; for the same reason, the use of these additives may prevent undesirable heating episodes (FAO 2001; Hardy 2010; Hardy and Roley 2000; National Organic Standards Board 2013).

In addition, the resistance of similar antioxidants against environmental conditions—and the consequent preservation of fish oil against rancidity—seem to vary season by season, with a suspected influence of collection season periods on the oxidative stability; storage conditions remain unchanged (Bragadóttir et al. 2004b; Syväoja et al. 1985).

The presence of metals acting as pro-oxidants (such as iron and copper) should be expected higher in fishmeal than in raw fish, when speaking of Nordic fish as capelin;

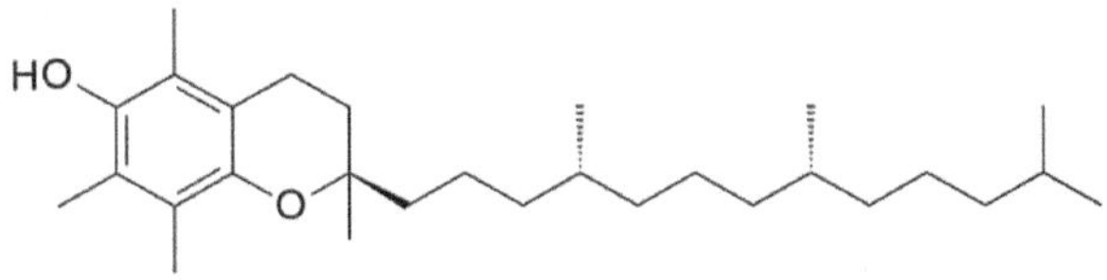

Fig. 1.1 Molecular structure of α-tocopherol, also named vitamin E, molecular formula: $C_{29}H_{50}O_2$, Chemical Abstracts Service number: 59-02-9, molecular weight: 430.717 Da. α-Tocopherol has been reported and studied recently when speaking of fishmeal supplementation in different ambits. It is considered a good antioxidant with reference to fishmeal stabilisation during shipping, with other natural mixtures such as rosemary extract, ethoxyquin, and/or butylated hydroxytoluene. BKChem version 0.13.0, 2009 (http://bkchem.zirael.org/index.html) has been used for drawing this structure

the same thing has been reported when speaking of peroxide values, iodine values, and polyene indexes (Bragadóttir et al. 2004b). The influence of heating processes cannot be excluded because the higher the possibility of heating episodes, the higher the degree of physicochemical modifications on fat molecules and protein at least. In detail, the production of free fatty acids by spontaneous hydrolysis has to be expected in these conditions, with concomitant low pH values.

Moreover, colorimetric modifications can be expected as a synergic function of seasonal collection periods and storage/processing conditions: interestingly, these chromatic variations (Parkers 1994) may be notably observed in all seasons except for spring, in relation to capelin. However, observed variations—also expressed as 'browning' (Pokorný et al. 1973)—may be in the acceptability range. Consequently, their importance could be not so important. With reference to ammonia, carbon monoxide, and sulphur dioxide, it may be inferred that observed variations during storage are mainly function of chemical and physical modifications ascribed to storage/processing conditions and fish diversity only, without a clear seasonal cause. Interestingly, carbon monoxide can be a powerful indicator in certain situations (Bragadóttir et al. 2004a; Ólafsdóttir et al. 1997). On a general level, observed variations in fish and fishmeal composition should be considered important in the first storage months. The most observed variations are (Prime 2018; Webster and Lim 2002):

(1) Degradation of fish protein with production of free amines (with correlated off-flavours and safety implications), possibility of cross-linking reactions (and consequent textural modifications, with important effects on fishmeal), and production of volatile compounds (in connection with obtained peroxides and free radicals after oxidation of fish oil).

(2) Decomposition of fatty molecules with production of free radicals, peroxides, fatty acids, and consequent colorimetric variations (catalysing agents: pro-oxidant metals, light oxygen availability). In addition, nitrogen-based molecules such as protein residues can easily react with the production of volatile compounds. The reduction of ω-3 fatty acids is notable, especially in fishmeal.

(3) Reduction of antioxidant natural substances such as vitamin A, α-tocopherol, astaxanthin (Fig. 1.2), and other compounds with similar action.

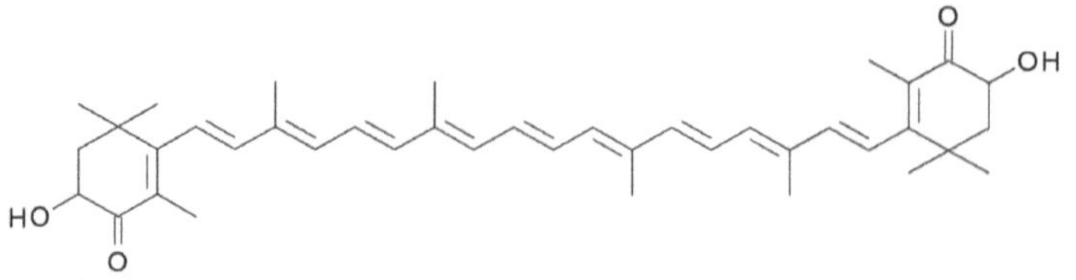

Fig. 1.2 Molecular structure of astaxanthin, molecular formula: $C_{40}H_{52}O_4$, molecular weight: 596.841 Da, an interesting antioxidant carotenoid found in many organisms including salmons and rainbow trouts. Astaxanthin and other antioxidants such as vitamin E (Fig. 1.1) are more abundant in the summer and spring seasons, while a remarkable diminution may be observed in winter and autumn, when speaking of certain fish types such as capelin. BKChem version 0.13.0, 2009 (http:// bkchem.zirael.org/index.html) has been used for drawing this structure

On the other side, non-natural contaminants such as polychlorinated biphenyls (PCB) and organochlorine insecticides (OI) should be taken into account and examined in fishmeal. In exclusive relation to certain fish species—anchovies (*Engraulis encrasicolus* L. 1758) and 'bonito' (*Sarda sarda* L. 1758)—in Turkey, it has been reported that PCB are present in both species, but anchovies were more contaminated than bonito; on the other hand, no OI traced have been found. This research (Cakirogullari and Secer 2011) demonstrated that different species could probably assume and act as vectors for industrial contaminants in different ways, depending on their nature. However, this problem is not correlated to seasonal reasons without a more profound analysis of PCB and OI utilisation throughout different seasons. Probably, these pesticides can influence seasonally obtained results depending on their use, similar to problems recently observed concerning honeybees (Chap. 4). In fact, detailed analyses on albacore tuna (*Thunnus alalunga*) concerning PCB and dichlorodiphenyltrichloroethane (DDT) have demonstrated the importance of collection locations—Reunion Island and South Africa—and the probably different levels of contamination (Munschy et al. 2016). The relation between environmental contamination and sure fish/fishmeal contamination is not assured, depending on many factors such as the position of selected fish species in the food web, and the natural differences between species' behaviours (Miniero et al. 2014; Paiano et al. 2013). In addition, the importance of precipitations cannot be excluded: it has been reported that the abundance of sodium pentachlorophenate in different freshwater fish species collected in the Dongting Lake (China) may temporally vary, in particular when comparing wet and dry seasons (Hu et al. 2018).

1.3   The Modern Seafood Industry and the Importance of Monitoring. Conclusions

Because of previously discussed arguments, it can be affirmed that the monitoring of fishmeal is extremely important when speaking of food production and related quality (Snyder et al. 1962). On the other side, routine controls should rely on affordable, easy-to-use, and rapid methods when speaking of the determination of different chemical parameters (Sect. 1.1). Consequently, the list of available analytical procedures could be long enough; however, the use of simple and rapid systems should be preferred when speaking of industrial plants. One of these examples is near infrared reflectance spectroscopy (NIRS) in relation to the reliable determination of moisture, oils, non-treated protein, total volatile nitrogen, and some other parameters useful for the determination of freshness and quality assessment. NIRS can be useful in many ambits, but the use for fishmeal control may solve many problems in a reasonable time, and with excellent results when speaking of crude fish and seafood, processed seafoods, non-fishmeal products (oysters, etc.), and the simple determination and discrimination of inedible fish parts such as fish bones (Cozzolino et al. 2002, 2005, 2009; Shen et al. 2017).

On the other hand, the possible substitution of fishmeal with soy proteins has to be considered because soy proteins have a good nutritional profile (Bjerkeng et al. 1997; Hossain and Koshio 2017; Vassallo-Agius et al. 2001a). Moist and soft-dry pellets have been considered as partial replacers for Japanese marine finfish (yellowtail, red seabream), but also for freshwater species such as carp and rainbow trout (Barclay et al. 2006; Lazo and Davis 2000; Vassallo-Agius et al. 2001b). It has been reported that soybean is up to 30% of the total amount of meals (Li et al. 2000; Ju et al. 2012); soybean is also one of the main ingredients (the list includes proteins, oil seeds, legumes, corn gluten meals, other cereals, and micro-algae). Consequently, the analytical profile of protein contents in fishmeal should take into account the replacement of fish-derived species such as brown fishmeal from tuna with soy proteins (Belal and Assem 1995; Elangovan and Shim 2000; Webster et al. 1992). Naturally, the higher the substitution, the lower the tendency to seasonal variations. Apparently, the trend appears the progressive substitution of fish-derived meal with vegetable sources. In these conditions, the low variability of used fishmeal should be expected in the next years.

Finally, the stabilisation of fishmeal with antioxidant preparations should be carefully discussed. At present, the European position by the European Food Safety Authority (EFSA) excludes the use of a peculiar substance, ethoxyquin (Fig. 1.3), for fishmeal preservation (European Commission 2017; Prime 2018). However, this additive and related mixtures with other substances may be placed on the market until 30 September 2019, provided that the use is clearly the incorporation in feed materials. In addition, feed materials containing ethoxyquin and related mixtures can be placed on the European market until 31 December 2019, while compound foods containing also this additive and/or related mixtures may be allowed on the market until 31 March 2020.

This situation forces fishmeal producers to find adequate countermeasures. At present, the following additives may be added to fishmeal preparations in the European Union with the complete exclusion of additional substances (Prime 2018):

(a) Ethoxyquin (50 ppm only)
(b) Butylated hydroxytoluene (100 ppm)

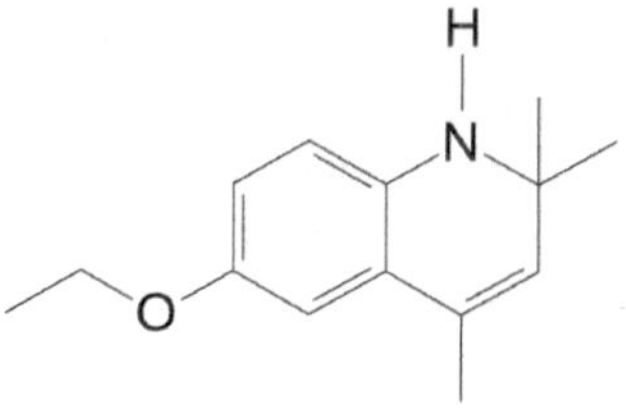

Fig. 1.3 Molecular structure of ethoxyquin, also named 6-ethoxy-2,2,4-trimethyl-1,2-dihydroquinoline, Chemical Abstracts Service number: 91-53-2, molecular formula: $C_{14}H_{19}NO$, molecular weight: 217.312 Da. This synthetic antioxidant may be placed on the market as an additive (and also mixed with other substances) until 30 September 2019, provided that the use is clearly the incorporation in feed materials. In addition, feed materials containing ethoxyquin and related mixtures can be placed on the European market until 31 December 2019, while compound foods containing also this additive and/or related mixtures may be allowed on the market until 31 March 2020. BKChem version 0.13.0, 2009 (http://bkchem.zirael.org/index.html) has been used for drawing this structure

(c) Tocopherol-based antioxidant (250 ppm), additive acronym: E306, while α-tocopherol is E307.

As a result, the analytical problem is relevant enough (International Maritime Organization 2017). With exclusive reference to oxidation tests, the following methods are available (Prime 2018):

(1) Oxidability test (one of these methods is the 'Rancimat' system)
(2) Oxygen consumption test (one of these methods is the 'Oxipres' method). This system measures oxygen pressure
(3) Accelerated ageing tests (autoxidation, photo-oxidation, chemical oxidation, or auto-oxidation procedures).

In relation to the present situation, and future development, it is expected that other testing methods have to concern:

(a) The amount of anisidine (this procedure evaluates only secondary oxidation products, including non-volatile molecules)
(b) The examination of peroxide values (this system concerns only the concentration of hydroperoxides in lipids as primary oxidation products
(c) Sensorial panels
(d) Hexanal (determination of volatile compounds as index of off-flavour production)
(e) The determination of thiobarbituric acid. This method concerns secondary oxidation products of PUFA.

References

Aquaculture Working Group (2012) Petition for listing on national list of approved and prohibited substances SEC. 2118. [7 U.S.C. 6517] National List. Tocopherols for Aquatic Animals, 27 Apr 2012. United States Department of Agriculture, Agricultural Marketing Service. Available https://www.ams.usda.gov/sites/default/files/media/Tocopherols%20%28Aquaculture%29.pdf. Accessed 06th Sept 2018

Astrup HN, Halvorsen JE (1985) Investigation on oxidation of feedstuffs. In: Marcuse R (ed) Lipid oxidation. Biological and food chemical aspects. Lipidforum proceedings, Scandinavian forum for lipid research and technology, SIK, Göteborg, pp 154–158

Barclay MC, Irvin SJ, Williams KC, Smith DM (2006) Comparison of diets for the tropical spiny lobster *Panulirus ornatus*: astaxanthin-supplemented feeds and mussel flesh. Aquacult Nutr 12(2):117–125. https://doi.org/10.1111/j.1365-2095.2006.00390.x

Barlow SM (1994) Is there a future for ethoxyquin and other synthetic antioxidants. In: Proceedings of the international fishmeal and oil manufacturers association annual conference, 29 Aug–2 Sept 1994, Copenhagen, pp 51–55

Belal IEH, Assem H (1995) Substitution of soybean meal and oil for fish meal in practical diets fed to channel catfish, *Ictalurus punctatus* (Rafinesque): effects on body composition. Aquacult Res 26(2):141–145. https://doi.org/10.1111/j.1365-2109.1995.tb00894.x

Bjerkeng B, Refstie S, Fjalestad KT, Storebakken T, Rødbotten M, Roem AJ (1997) Quality parameters of the flesh of Atlantic salmon (*Salmo salar*) as affected by dietary fat content and full-fat soybean meal as a partial substitute for fish meal in the diet. Aquacult 157(3–4):297–309. https://doi.org/10.1016/s0044-8486(97)00162-2

Boran G, Boran M, Karaçam H (2008) Seasonal changes in proximate composition of anchovy and storage stability of anchovy oil. J Food Qual 31(4):503–513. https://doi.org/10.1111/j.1745-4557.2008.00215.x

Bragadóttir M, Pálmadóttir H, Kristbergsson K (2004a) Composition and chemical changes during storage of fish meal from capelin (*Mallotus villosus*). J Agric Food Chem 52(6):1572–1580. https://doi.org/10.1021/jf034677s

Bragadóttir M, Pálmadóttir H, Kristbergsson K (2004b) Effect of fish meal processing on endogenous anti-and prooxidants in capelin (*Mallotus villosus*). J Agric Food Chem 52(6):1572–1580. https://doi.org/10.1021/jf034677s

Buttle LG, Crampton VO, Willams PD (2001) The effect of feed pigment type on flesh pigment deposition and color in farmed Atlantic salmon, *Salmo salar*. Aquacult Res 32(2):103–111. https://doi.org/10.1046/j.1365-2109.2001.00536.x

Cakirogullari GC, Secer S (2011) Seasonal variation of organochlorine contaminants in bonito (*Sarda sarda* L. 1758) and anchovy (*Engraulis encrasicolus* L. 1758) in Black Sea region, Turkey. Chemosph 85(11):1713–1718. https://doi.org/10.1016/j.chemosphere.2011.09.017

Cozzolino D, Chree A, Murray I, Scaife JR (2002) The assessment of the chemical composition of fishmeal by near infrared reflectance spectroscopy. Aquacult Nutr 8(2):149–155. https://doi.org/10.1046/j.1365-2095.2002.00206.x

Cozzolino D, Murray I, Chree A, Scaife JR (2005) Multivariate determination of free fatty acids and moisture in fish oils by partial least-squares regression and near-infrared spectroscopy. LWT Food Sci Technol 38(8):821–828. https://doi.org/10.1016/j.lwt.2004.10.007

Cozzolino D, Chree A, Murray I, Scaife JR (2009) Usefulness of near infrared spectroscopy to monitor the extent of heat treatment in fish meal. Int J Food Sci Technol 44(8):1579–1584. https://doi.org/10.1111/j.1365-2621.2008.01845.x

Dersjant-Li Y (2002) The use of soy protein in aquafeeds. In: Cruz-Suárez LE, Ricque-Marie D, Tapia-Salazar M, Gaxiola-Cortés MG, Simoes N (eds) Avances en Nutrición Acuícola VI. Memorias del VI Simposium Internacional de Nutrición Acuícola, 3–6 Sept 2002, Cancún, Quintana Roo, México. Available http://universidad.uanl.mx/utilerias/nutricion_acuicola/VI/archivos/A34.pdf. Accessed 06th Sept 2018

Drew MD, Ogunkoya AE, Janz DM, Van Kessel AG (2007) Dietary influence of replacing fish meal and oil with canola protein concentrate and vegetable oils on growth performance, fatty acid composition and organochlorine residues in rainbow trout (*Oncorhynchus mykiss*). Aquacult 267(1–4):260–268. https://doi.org/10.1016/j.aquaculture.2007.01.002

Elangovan A, Shim KF (2000) The influence of replacing fish meal partially in the diet with soybean meal on growth and body composition of juvenile tin foil barb (*Barbodes altus*). Aquacult 189(1–2):133–144. https://doi.org/10.1016/s0044-8486(00)00365-3

European Commission (2017) Commission implementing regulation (EU) 2017/962 of 7 June 2017 suspending the authorisation of ethoxyquin as a feed additive for all animal species and categories. Off J Eur Union L145:13–17

FAO (2001) Aquaculture development—1. Good aquaculture feed manufacturing practice. FAO Technical Guidelines for Responsible Fisheries 5, Suppl. 1. Food and Agriculture Organization of the United Nations, Rome. Available http://www.fao.org/3/a-y1453e.pdf. Accessed 06th Sept 2018

Hardy RW (2010) Utilization of plant proteins in fish diets: effects of global demand and supplies of fishmeal. Aquacult Res 41(5):770–776. https://doi.org/10.1111/j.1365-2109.2009.02349.x

Hardy RW, Roley DD (2000) Lipid oxidation and antioxidants. In: Stickney RR (ed) Encyclopedia of aquaculture. Wiley Inc., New York, pp 470–476

Hossain MS, Koshio S (2017) Dietary substitution of fishmeal by alternative protein with guanosine monophosphate supplementation influences growth, digestibility, blood chemistry profile, immunity, and stress resistance of red sea bream, *Pagrus major*. Fish Physiol Biochem 43(6):1629–1644. https://doi.org/10.1007/s10695-017-0398-4

Hu Y, Yi C, Li J, Shang X, Li Z, Yin X, Chen B, Zhou Y, Zhang Y, Wu Y (2018) Correction to: seasonal variations of PCDD/Fs in fishes: inferring a hidden exposure route from Na-PCP application for schistosomiasis control. Environ Monit Assess 190(6):190–331. https://doi.org/10.1007/s10661-018-6706-3

Hultin HO (1992) Lipid oxidation in fish muscle. In: Flick GJ, Martin RE (eds) Advances in seafood biochemistry composition and quality. Technomic Publishing, Lancaster, pp 99–122

International Maritime Organization (2017) Amendments to the IMDG code and supplements—proposed amendment to the shipping provisions for FISHMEAL (FISHSCRAP), STABILIZED (UN 2216). CCC 4/6/14, 7th July 2017. Sub-committee on Carriage of Cargoes and Containers, International Maritime Organization, London

Jowett WG, Davies W (1938) A chemical study of some Australian fish. CSIRO pamphlet 85. HJ Green, Government Printer, Melbourne

Ju ZY, Deng DF, Dominy W (2012) A defatted microalgae (*Haematococcus pluvialis*) meal as a protein ingredient to partially replace fishmeal in diets of Pacific white shrimp (*Litopenaeus vannamei*, Boone, 1931). Aquacult 354–355:50–55. https://doi.org/10.1016/j.aquaculture.2012.04.028

Lazo JP, Davis D (2000) Ingredients and feed evaluation. In: Stickney RR (ed) Enclyclopedia of aquaculture. Wiley Inc., New York, pp 453–463

Li MH, Robinson EH, Hardy RW (2000) Protein sources for feeds. In: Stickney RR (ed) Enclyclopedia of aquaculture. Wiley Inc., New York, pp 688–695

McBride JR, MacLeod RA, Idler DR (1959) Proximate analysis of Pacific herring (*Clupea pallasii*) and an evaluation of tester's 'fat factor. J Fish Res Board Can 16(5):679–684. https://doi.org/10.1139/f59-049

Miniero R, Abate V, Brambilla G, Davoli E, De Felip E, De Filippis SP, Dellatte E, De Luca S, Fanelli R, Fattore E, Ferri F, Fochi I, Fulgenzi AR, Iacovella N, Iamiceli AL, Lucchetti D, Melotti P, Moret I, Piazza R, Roncarati A, Ubaldi A, Zambon S, di Domenico A (2014) Persistent toxic substances in mediterranean aquatic species. Sci Total Environ 494–495:18–27. https://doi.org/10.1016/j.scitotenv.2014.05.131

Munschy C, Bodin N, Potier M, Héas-Moisan K, Pollono C, Degroote M, West W, Hollanda SJ, Puech A, Bourjea J, Nikolic N (2016) Persistent organic pollutants in albacore tuna (*Thunnus alalunga*) from reunion island (Southwest Indian Ocean) and South Africa in relation to biological and trophic characteristics. Environ Res 148:196–206. https://doi.org/10.1016/j.envres.2016.03. 042

National Organic Standards Board (2013) Petitioned material proposal—tocopherols aquaculture. National Organic Standards Board, Livestock Subcommittee, 22 Aug 2013, Reviewed and revised 21 Jan 2014. United States Department of Agriculture, Agricultural Marketing Service. Available https://www.ams.usda.gov/sites/default/files/media/tocopherols%20aqua% 20proposal%202014.pdf. Accessed 06th Sept 2018

Nickell D, Bromage NR (1998) The effect of dietary level on variation of flesh pigmentation in rainbow trout (Oncorhynhus mykiss). Aquacult 161(1–4):237–251. https://doi.org/10.1016/ s0044-8486(97)00273-1

Ólafsdóttir G, Hognadóttir AA, Martinsdóttir E (1997) Application of gas sensors to evaluate freshness and spoilage of various seafoods. In: Ólafsdóttir G, Luten J, Dalgaard P, Careche M, Verrez-Bagnis V, Martinsdóttir E, Heia K (eds) Methods to determine the freshness of fish in research and industry. International Institute of Refrigeration, Paris, pp 100–109

Ólafsson P (1953) Some observations on extraction and iodine values of fat from fish tissue, press cake and meal, and on peroxide values of meal fat the first day. Timarit Verkfraedingafelags Isl 38(4):102–108

Opstvedt J (1975) Influence of residual lipids on the nutritive value of fish meal. VII. Effect of lipid oxidation on protein quality of fish meal. Acta Agric Scand 25(1):53–71. https://doi.org/10.1080/ 00015127509435036

Opstvedt J (1985) Fish lipids in animal nutrition. International association of fish meal manufacturers (IAFMM) Tech Bull 22:1–27

Paiano V, Generoso C, Mandich A, Traversi I, Palmiotto M, Bagnati R, Colombo A, Davoli E, Fanelli R, Fattore E (2013) Persistent organic pollutants in sea bass (Dicentrarchus labrax L.) in two fish farms in the Mediterranean Sea. Chemosph 93(2):338–343. https://doi.org/10.1016/j. chemosphere.2013.04.086

Parkers RW (1994) Measurement of colour in food. Food Technol Int Eur 175–176

Pike IH, Andorsdóttir G, Mundheim H (1990) The role of fish meal in diets for salmonids. IAFMM Tech Bull 24:1–35

Pokorný J, El-Zeany BA, Janíček G (1973) Non-enzymic browning III: browning reactions during heating of fish oil fatty acid esters with protein. Z Lebensm Unters Forsch 151(1):31–35. https:// doi.org/10.1007/bf01384278

Prime D (2018) ANTIOXIDANTS—fish meal and Feed. Vitablend Protection Business Unit, Wolvega. Available http://sfs.is/wp-content/uploads/2018/05/Antioxidants-Vitablend-Dvid-Prime.pdf. Accessed 06th Sept 2018

Romoser GL, Dudley WA, Burke RP (1968) Antioxidants in fish meal. Fish News Int 1:27–29

Shen G, Han L, Fan X, Liu X, Cao Y, Yang Z (2017) Classification of fish meal produced in China and Peru by online near infrared spectroscopy with characteristic wavelength variables. J Near Infrared Spectrosc 25(1):63–71. https://doi.org/10.1177/0967033516686041

Snyder DG, Ousterhout LE, Titus HW, Morgareidge K, Kellenbarger S (1962) The evaluation of the nutritive content of fish meals by chemical methods. Poult Sci 41(6):1736–1740. https://doi. org/10.3382/ps.0411736

Syväoja EL, Salminen K, Piironen V, Varo P, Kerojoki O, Koivistoinen P (1985) Tocopherols and tocotrienols in finnish foods: fish and fish products. J Am Oil Chem Soc 62(8):1245–1248. https:// doi.org/10.1007/bf02541835

Tolasa S, Cakli S, Ostermeyer U (2005) Determination of astaxanthin and canthaxanthin in salmonid. Eur Food Res Technol 221(6):787–791. https://doi.org/10.1007/s00217-005-0071-5

Undeland I (1997) Lipid oxidation in fishscauses, changes and measurements. In: Ólafsdóttir G, Luten J, Dalgaard P, Careche M, Verrez-Bagnis V, Martinsdóttir E, Heia K (eds) Methods to determine the freshness of fish in research and industry. International Institute of Refrigeration, Paris, pp 241–257

UNECE (2016) Addendum to ST/SG/AC.10/C.3/2016/82: Special provision for fish meal (Fish Scrap), Stabilised (UN 2216): Class 9. Committee of Experts on the transport of dangerous goods and on the globally harmonized system of classification and labelling of chemicals—sub-committee of experts on the transport of dangerous goods, UN/SCETDG/50/INF.24, 14th Nov 2016. United Nations Economic Commission for Europe and Executive Committee (UNECE), Geneva

Vassallo-Agius R, Watanabe T, Imaizumi H, Yamazaki T, Satoh S, Kiron V (2001a) Effects of dry pellets containing astaxanthin and squid meal on the spawning performance of striped jack *Pseudocaranx dentex*. Fish Sci 6–7(4):667–674. https://doi.org/10.1046/j.1444-2906.2001.00304.x

Vassallo-Agius R, Imaizumi H, Watanabe T, Yamazaki T, Satoh S, Kiron V (2001b) The influence of astaxanthin-supplemented dry pellets on spawning of striped jack. Fish Sci 67(2):260–270. https://doi.org/10.1046/j.1444-2906.2001.00248.x

Waissbluth MD, Guzman L, Plachco FP (1971) Oxidation of lipids in fish meal. J Am Oil Chem Soc 48(8):373–424. https://doi.org/10.1007/bf02637366

Wang Y, Kong LJ, Li C, Bureau DP (2006) Effect of replacing fish meal with soybean meal on growth, feed utilization and carcass composition of cuneate drum (*Nibea miichthioides*). Aquacult 261(4):1307–1313. https://doi.org/10.1016/j.aquaculture.2006.08.045

Webb JE, Brunson CC, Yates JD (1973) Effects of feeding fish meal and tocopherol on flavor of precooked. Frozen Turkey Meat Poult Sci 52(3):1029–1034. https://doi.org/10.3382/ps.0521029

Webster CD, Lim C (2002) Nutrient requirements and feeding of finfish for aquaculture. C.A.B. International, Wallingford

Webster CD, Yancey DH, Tidwell JH (1992) Effect of partially or totally replacing fish meal with soybean meal on growth of blue catfish (*Ictalurus furcatus*). Aquacult 103(2):141–152. https://doi.org/10.1016/0044-8486(92)90408-d

Chapter 2
Water, Carbon, and Phosphorus Footprint Concerns in the Food Industry

Abstract This chapter discusses recent environmental concerns of the food consumer. The expansion of human populations worldwide and other factors have progressively caused the continuous increase of lands for agricultural purposes, with consequent deforestation in developing areas above all. At the same time, the trend of produced foods per area of agricultural land has been constantly increased with the consequent enhancement of the growth of human beings, although the amount of cultivated lands has been reduced in the last forty years. Food production (or food overproduction) is often correlated with environmental concerns because of the main role of three resources and their increasing consumption by food industries: water, energy, and anti-pest agents. More than a single food or food category may be interested in this way. For these reasons, three peculiar indicators—water footprint, carbon footprint (concerning the energetic consumption), and phosphorus footprint (with reference to pesticides and other similar chemical compounds)—are popular at present. However, are these footprints good indicators? This chapter discusses advantages and risks associated with these variables by a broader perspective.

Keywords Carbon footprint · Global warming · Greenhouse gas emission · Pesticide · Phosphorus footprint · Pollution · Water footprint

Abbreviations

CO_2e	Carbon dioxide equivalent
CF	Carbon footprint
FAO	Food and Agriculture Organization of the United Nations
GC	Gas chromatography
GHG	Greenhouse gas
CH_4	Methane
N_2O	Nitrous oxide
NGO	Non-governmental organisation
P	Phosphorus
PF	Phosphorus footprint

© The Author(s), under exclusive license to Springer Nature Switzerland AG 2019
S. D. Sharma et al., *Raw Material Scarcity and Overproduction in the Food Industry*, Chemistry of Foods, https://doi.org/10.1007/978-3-030-14651-1_2

RSPO   Roundtable on Sustainable Palm Oil
UK   United Kingdom
WF   Water footprint

2.1   Food Agriculture and Environmental Concerns

At present, the expansion of human populations worldwide is one of the main causes—and the main effect also—of the continuous increase of lands for agricultural
purposes, with consequent deforestation in developing areas above all (Barbier 2004;
Ewert et al. 2005; Matson et al. 1997; Rosegrant et al. 2001). At the same time, the
trend of produced foods per area of agricultural land has been constantly increased
(Grigg 1993; Meyer and Turner 1992; Naylor 1996; Rounsevell et al. 2003) with the
consequent enhancement of the growth of human beings, although the amount of
cultivated lands has been reduced in the last forty years.

Reasons for this apparently contradictory behaviour in the current agricultural
productions are not easily examinable because of their intrinsic variety. Biological,
physical, and socio-economic features should be evaluated and critically discussed in
a complete and synergic ambit (Easterling et al. 2001; Ewert et al. 2005). Interestingly,
similar studies can be used to predict the future of crop production in the next decades
(Ewert et al. 2005; Rounsevell et al. 2003).

In particular, the following variables should be considered (Downing et al. 1999;
Ewert et al. 2005; Harrison and Butterfield 1996; Jamieson et al. 1999; Landau et al.
1998; Parry et al. 2004; Rosenzweig and Parry 1994; Rounsevell et al. 2003):

(a) Climate modifications and abundance of atmospheric carbon dioxide
(b) Differences between regional and local climates
(c) Seasonal differences, related to both climates and plants
(d) Scarcity (or abundance) of bioavailable water, including the precipitations
(e) Scarcity (or abundancy) of bioavailable nutrients
(f) Amount of used pesticides, insecticides, and so on
(g) Influence of pests and various diseases attacking vegetable species
(h) Influence of pollination (Chap. 3)
(i) Soil features (acidity, salinity, etc.)
(j) Evolution in agricultural and breeding techniques, with possible local variations
(k) Other anthropic reasons (Chap. 3).

In should be noted, with reference to the evolution in agricultural and breeding techniques (with possible local variations), that the use of pesticides and other
substances against non-useful vegetable species should be considered in this ambit.
At the same time, agricultural training for farmers is critical. In addition, 'breeding'
amelioration may concern the use of more resistant, more productive, and more stress-
resistant plants in comparison with 'normal' cultivar types. This point is extremely
important because obtained yields depend mainly on the joint availability of 'powerful' soils (and correlated 'abundance of nutrients') on the one side, of excellent

plants with enhanced and demonstrated yielding capability, and finally of available carbon dioxide, even if the effect of this molecule on real yield capability could be questioned. In fact, the positive effect of increasing carbon dioxide is reported to be demonstrable when speaking of potential yields, by means of dedicated process-based models (Amthor 1998; Boote et al. 1997; Ewert et al. 2005; Tubiello and Ewert 2002). With reference to yield productions correlated to several species, cereals are the best choice so far and the reference crop (wheat). Consequently, species such as *Triticum aestivum* are surely preferred (Ewert et al. 2005) if compared with *Solanum tuberosum* (potatoes). On the other side, it should be noted that the continuous demand for agricultural land—actual yields in developing countries have been reported to reach 80% so far (Oerke and Dehne 1997)—has often concerned also urban areas. These lands can be improved mainly by means of innovative technological instruments (Austin 1999; Evans and Fischer 1999; Ewert et al. 2005; Johnson 1999; Reynolds et al. 1999). Other general factors influencing indirectly production yields—the real influenced parameter appears the concentration of atmospheric carbon dioxide—may be increasing temperature values, nitrogen availability, and water abundance (Kimball et al. 2002; Long 1991; Morison and Lawlor 1999).

Anyway, food production—better defined 'food overproduction'—is often correlated with environmental concerns because of the main role of three resources and their increasing consumption by food industries:

(a) Water
(b) Energy, in terms or used energy sources (carbon, fuel, hydroelectric energy, solar energy, etc.)
(c) Anti-pest agents.

Each of these points should be considered carefully when speaking of environmental impacts and consequent damages, including climate variations (Ridoutt et al. 2011; UN News 2007). By the social viewpoint, each activity able to increase its own amount of used water, energy, or pesticides should be considered with suspect. The demonstration of lower consumptions in these specific ambits—especially with concern to food production sectors—would be highly appreciated, and probably publicised by means of adequate labelled messages (Boardman 2008; Galli et al. 2012; Hoekstra 2013; Leach et al. 2016; Segal and MacMillan 2009). More than a single food or food category may be interested in this way. One of the most known and recognised organisations concerning the production of environmentally sustainable products is the Roundtable on Sustainable Palm Oil (RSPO): this association includes many stakeholders involved and/or interested in the production and the promotion of use of sustainable palm oil, provided that several conditions are respected, including also[1]:

(a) The protection of biodiversity
(b) The involved ecosystems are enhanced.
(c) An efficient planning of the use of agricultural lands is assured.

---

[1] More info can be accessed at the following website: https://rspo.org/.

As an example, this and other organisations have considered peculiar objectives such as the diminution of environmental pollution and the correlated decrease of greenhouse gas (GHG) emissions. Consequently, an inverse proportion can be obtained correlating environmental damages to lands and natural habitats on the one side, and a selected group of chemical or physical indicators. Because of the virtually identical importance of water, energy, and pest-control agents in many food production environments, the attention of food and environmental stakeholders, including many non-governmental organisations (NGO), has been progressively focused on three peculiar indicators, also named 'footprints' (Hoekstra et al. 2011; Chapagain et al. 2006; Hoekstra and Mekonnen 2012; Steen-Olsen et al. 2012):

(a) Water footprint
(b) Carbon footprint (concerning the energetic consumption)
(c) Phosphorus footprint (with reference to pesticides and other similar chemical compounds).

These rising trends can have environmental effects, and food productions may be negatively affected. The three following sections aim to discuss each of these indicators separately, although the synergistic action of different footprints has to be expected. As a little premise, it should be noted that the word 'footprint' can suggest the pressure of a well-defined indicator on land and environmental health because of the possible expression in area-based units, such as the so-called Ecological Footprint measured in hectares (Wiedmann and Minx 2008). However, this chapter discusses all footprints in chemical terms, without relation to area-based definitions, or similar proposals.

2.2 Water Use in Agriculture, Scarcity, and Correlated Results

The use of water in the modern agriculture has to be carefully considered. Consequently, the term 'water footprint' (WF) has progressively been used in environmental and economic studies. It can be affirmed that WF concerns the amount of needed freshwater for agricultural and other purposes (anthropic activities), and it is normally measured in m^3. It should also be clarified that WF concerns three different freshwater types (Mekonnen and Hoekstra 2011; Steen-Olsen et al. 2012):

(1) Green water. It corresponds to the quantity of water obtained by precipitation (rainwater) directly used by plants.
(2) Blue water. This term means superficial and ground water only. In general, this water should not be mentioned in this ambit, although the continuous decrease of green water for plants indirectly means that blue water is forced to decline.
(3) Grey water. This water has to be considered as the amount of freshwater needed to dilute the concentration of pollutants. In general, this water should not be taken into account when speaking of water consumption.

WF data may be very useful when speaking of the contribution to environmental damages by one specific industry, business, or individual entity (household consumption). However, the most part of researches concern WF data related to specified geographical areas: national regions or entire countries (Hoekstra and Mekonnen 2012). Anyway, correct WF data should refer mainly to green water and/or blue water (because of its dependency from availability of green water).

WF data can be very variegated in the most developed countries and in developing countries also. It has been reported that WF in Australian areas may not be excessive (Ridoutt et al. 2011) if compared with 'carbon footprinting' (Sect. 2.3). This apparent conclusion depends on the normalised 'weight' of carbon footprint on a broad scale when speaking of massive livestock activities in Australia; on the other side, it has been supposed that the true reason is the negligible importance of water decrease in these areas because of the supposed high water availability. In relation to the European Union, several WF data suggest that the most important effects may be observed in the Mediterranean regions for climatic reasons (high summer temperatures, lack of springs, etc.). In fact, WF values concerning blue water only (year: 2004) appear high in the Mediterranean area because of scarce rains, high temperatures, and the prevailing importance of irrigation for agricultural purposes (maximum WF value: 438 m^3 per capita, Spain; average value in the EU: 179 m^3 per capita; minimum quantity: 39 m^3 per capita, Finland). These results (FAOSTAT 2018; Steen-Olsen et al. 2012) may be simplified as shown in Fig. 2.1, where:

(a) The following countries—Spain, Cyprus, Luxembourg, and Greece—exceed 350 m^3 per capita. With the exception of Luxembourg, these countries belong to the Mediterranean area (advanced agricultural activities, irrigation-dependent agriculture, scarce precipitations, reduced access to freshwater).

(b) WF values between 200 and 350 m^3 per capita concern only Portugal, Belgium, and Italy. With the exception of Belgium, these countries belong to the Mediterranean area (advanced agricultural activities, irrigation-dependent agriculture, scarce precipitations, reduced access to freshwater).

(c) WF values between 100 and 200 m^3 per capita are ascribed to many countries: France, Netherlands, UK, Denmark, Ireland, Germany, Austria, Finland, Sweden, Malta, and Slovenia. Interestingly, these EU Member States are all developed countries and belonging to Nordic or Atlantic areas, with the exception of Malta (little dimensions) and Slovenia.

(d) Finally, WF is reported ≤100 m^3 per capita in relation to the following EU countries only: Hungary, Romania, Estonia, Czech Republic, Slovakia, Latvia, Bulgaria, Lithuania, and Poland. Substantially, these EU countries belong to East Europe (reduced industrialisation, high water availability, and good precipitations).

In general, WF data related to the entire EU area are not excessive if compared with global data (FAOSTAT 2018; Steen-Olsen et al. 2012). The main result of these investigations—probably verifiable in other areas—is that the higher the dependency on intensive agriculture, the higher the bluewater consumption (and the lower the availability of rainwater). Consequently, it appears that WF can be a useful indicator

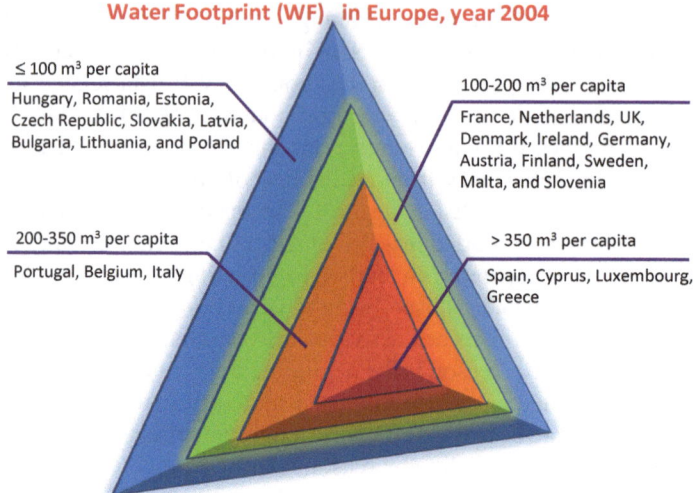

Fig. 2.1 WF data can be very variegated. In relation to the apparently low importance of WF in comparison with 'carbon footprinting', it has been supposed that the true reason is the negligible importance of water decrease in these areas because of the supposed high water availability. In relation to the European Union, several WF data concerning blue water only for the year 2004 (FAOSTAT 2018) suggest that the most important effects may be observed in the Mediterranean regions for climatic reasons because of scarce rains, high temperatures, and the prevailing importance of irrigation for agricultural purposes. These results, in terms of m^3 per capita, may be simplified and correlated with European countries

in certain ambits, when the parameterisation and normalisation of this indicator are considered in a broad situation with other different indexes (Sect. 2.3). However, the dependence on geographical localisation, the importance of several anthropic activities, and the climate modifications have to be seriously taken into account before stating conclusions (Pfister et al. 2009; Steen-Olsen et al. 2012).

2.3 Carbon Footprint. Is It a Good Indicator?

Carbon footprint (CF) may be defined in various ways. For example, it may be considered as the specific indicator for global climate change (also defined 'global warming'); in this ambit, it can be measured as carbon dioxide equivalents (CO_2e) (Hertwich and Peters 2009; Pandey et al. 2011; Peters 2010; Ridoutt et al. 2011; Wiedmann and Minx 2008). More specifically, it may be also defined as the total quantity of carbon dioxide and other GHG that can be really emitted over the complete life cycle of a product, or a process, according to the United Kingdom (UK) Parliamentary Office of Science and Technology (POST 2006). Consequently, it can be measured in grams of CO_2e per kilowatt-hour of generation, globally representing carbon dioxide, methane, and nitrous dioxide at least, taking also into account that (Capper 2011):

(1) The 'global warming potential' of carbon dioxide can be assumed $= 1$.
(2) The 'global warming potential' of methane can be assumed $= 25$.
(3) The 'global warming potential' of nitrous dioxide can be assumed $= 298$.

This definition is conveniently broadened. On the other side, CF may be also expressed as (Wiedmann and Minx 2008):

(1) The quantity of carbon dioxide emitted as the result of the combustion of fossil fuels by a small organisation or industry (Grubb and Ellis 2007), or
(2) The measure of anthropic activities on the environments in terms of produced GHG, provided that these amounts are measures as tons of carbon dioxide.

Actually, many definitions may be expressed, but the meaning is always correlated to the production of energy by means of the combustion of energy sources, and the result can be considered in terms of produced carbon dioxide. Interestingly, one of the main differences between available definitions is always the specification of emitting activities—the complete world of industries, or a selected subsector, or a single unit (a food company, a commodity reaching market, and so on). Another important difference may concern the environmental impact as defined for a selected amount of products, or for a defined and temporally restricted process (Wiedmann and Minx 2008).

Another concept should be considered when speaking of CF: differently from some possible opinion, this indicator cannot represent the complete environmental impact in research studies (Laurent et al. 2010, 2012; Rees 1996), or be a reliable index concerning environmental burden, as intended by the norm ISO 14040: 2006 (ISO 2006). Actually, CF is largely considered because of its easy and reasonable measurement: the simple CF determination requires only two data, produced electricity and used fuel, in many ambits (Ridoutt et al. 2011). Generally, these data are considered when speaking of entities such as (Anonymous 2006a; Trucost 2006; Wiedmann and Minx 2008):

(1) Separated countries
(2) Regional areas within national boundaries
(3) Communities (schools, public institutions, etc.)
(4) Food and non-food products and/or services.

In other terms, the analysis of emissions should take into account:

(a) The involved activities producing carbon dioxide
(b) The final product(s) or service(s).

It has been also reported that the localisation of CO_2e-producing activities is important: on-site activities should be taken into account separately, and off-site activities would be evaluated only as indirect CF causes. As an example, the CF index in UK was evaluated in 2001 for household communities as the sum of different portions, including (Anonymous 2006b; DEFRA 2006):

(1) Direct (household) fuel consumption
(2) Household consumption of electricity
(3) Other household emissions
(4) Traffic-related causes (private cars)
(5) Public and private transportation systems
(6) Recreation, tourism, etc.
(7) Health and sanitisation services
(8) Food consumption (this point should take also into account catering services)
(9) Clothing and footwear
(10) Other products and/or services.

Substantially, the CF calculation should be related to the following 'emitting agents' at least:

(1) On-site emission agents: direct fuel use for different reasons, including electricity production
(2) Off-site emission activities: transportation, supply, and use of electricity produced off-site
(3) Off-site transportation activities (public transport, aviation, etc.) related to the human user
(4) Use of chemicals, furniture, food- and non-food products and services (including also transportation activities for foods, non-food articles, living animals, etc.).

Naturally, the list is not exhaustive (a simple example: the use of foods and beverages should take into account the possible use of catering services). However, this example could be useful to understand the complexity of the problem (and the real importance of CF index alone). Interestingly, the indirect use of electricity and collateral (off-site) emissions may account for 70% of the total CF in UK, while transportation activities—often considered as the worst environmental damage—should not exceed 28% (Wiedmann and Minx 2008). Different countries may show different results, and the problem of 'double counts' (the CF estimation may be sometimes questionable because of the repeated attribution of certain data to more than a single CF agent) is always possible (Hammerschlag and Barbour 2003; Lenzen 2008; Lenzen et al. 2007).

By the global viewpoint, it has to be considered that the main CF concerns come from the increasing use of fossil fuel combustions, generating a notable CO_2e portion (28.6%), according to the Intergovernmental Panel on Climate Change. Other gases, in accordance with the Kyoto Protocol (UN 1998), include methane (CH_4) and nitrous oxide (N_2O); these GHGs are reported to reach 14.3 and 7.9%, respectively, of the global CF (Core Writing Team et al. 2007; Pandey et al. 2011). Unfortunately, the last two gases are mainly correlated with agricultural activities. Consequently, the public opinion has progressively been concerned with reference to global warming and climate changes in terms of CF augment as the direct or indirect consequence of agricultural sectors. However, available information are often questionable and dependent on used calculations: these estimations depend mainly on available guidelines, but a plethora of possible choices makes impossible the critical discussion and comparison of different studies and researches (Kenny and Gray 2008; Padgett et al. 2008; Pandey et al. 2011; Schiermeier 2006; Wiedmann and Minx 2008). With exclusive reference to CF contribution derived by agricultural activities, one of the best strategies when calculating CF should be the use of a statistical database (FAOSTAT 2018) concerning agricultural activities by the Food and Agriculture Organization of the United Nations (FAO): the FAOSTAT. In this way, it could be possible to calculate most recent emissions associated with agricultural activities, waste production, and the associated transportation activities of vegetables, animals, and so on, until the final production of food and beverage items. These statistic calculations concern the CF contribution associated with carbon dioxide, N_2O, and CH_4. Interestingly, it has been reported that carbon dioxide contribution cannot be sufficient in this ambit (Pandey et al. 2011). Anyway, emission factors are nowadays available for many countries and regions, including compared data at the global level (Pandey et al. 2011).

By the analytical viewpoint, collected data can also derive from real-time analyses, for verification purposes. These ground-based analytical methods, concerning carbon dioxide and other volatile compounds, should include optical methods, the use of biosensors or chemical instruments able to collect and instantly evaluate the concentration of the analyte by means of infrared rays (concerning carbon dioxide), or gas chromatography (GC) when speaking of various analytes (Berg et al. 2006; USCCTP 2005). Naturally, the collection of gases concerning a notable land extension should be performed by means of flux towers, eddy covariance, and cavity ring-down spectrometers (Velasco et al. 2005). The problem is that obtained and calculated data should be turned into CO_2e by means of dedicated conversion factors; however, the lack of uniformity between guidelines, different researches, the intrinsic difference between regional areas, and the availability of user-friendly online calculators have probably complicated the analysis of the CF problem. In addition, CF could be reason for money transactions based on more or less acceptable results; as a result, CF may potentially influence business on a global level. Consequently, the use of shared and official guidelines should be needed (Pandey et al. 2011). At the same time, the comparison between CF and WF is difficult, even if the contribution of CF in the global assessment of environmental damages (eco-sustainability) can represent more than 90% if compared with WF in agricultural activities. In the ambit of animal production systems. It has to be noted that (Ridoutt et al. 2011):

(1)   Livestock enteric fermentation is responsible for emissions, and the name thing has to be affirmed for manure and produced urine.
(2)   With reference to soils, the use of inorganic nitrogen fertilisers and the residuation of cultivated leguminous pastures have to be considered.
(3)   The deforestation should be taken into account.
(4)   Other factors concern the role of fuel consumption, electricity, other fertilisers, other pasture materials, veterinary consultancies, and different services (including transportation).

The aggregated analysis of CF and WF (or CF alone) should be performed by using one of the most known life cycle impact assessment methodologies, such as the

PAS 2050:2008 system proposed by the British Standard Institution (BSI 2008). The discussion of similar methodologies is not among the scopes of this book; however, it can be affirmed here that this and other procedures are needed when speaking of comparing different footprint indexes because of their intrinsic difference (CF concerns only gaseous emissions, while WF concerns only the impact of water overuse). Anyway, results should be given and evaluated in terms of produced damages to human health, to ecosystem quality, and to resources as the result of carbon dioxide production and water use. The expression can be on a daily, monthly, or yearly basis. Consequently, CF or WF alone have to be parameterised and normalised in the broad ambit of a life cycle assessment procedure; simple CF or WF values may be not helpful when speaking of environmental burning, damages, and so on (Laurent et al. 2012).

The problem of GHG emissions has also forced the scientific and technological word to find some possible renewable sources with the aim of reducing the dependency on fossil fuels and the amount of produced carbon dioxide. One of the most known and debated solutions is ethanol fuel, often dubbed synonym of 'green energy'. However, the effects of the total or partial replacement of 'normal' energy sources with ethanol (it may be used with gasoline in the so-called gasohol mixture) should be evaluated in a broad perspective, without the exclusion of indirect (off-site) emission activities (de Oliveira et al. 2005). In addition, it has been recently reported that there are no distinctive advantages—in terms of reduced energetic consumption and carbon dioxide production—when speaking of partial or total fuel substitution with ethanol. On the other side, it has been suggested that the production of ethanol by sugarcane plants (Brazil) or corn plants (USA)—the whole production cycle concerns growing, harvesting, and biomass conversion to ethyl alcohol—is not favourable when speaking of ecological damages. On the contrary, the forestation—or the diminution of agricultural land—would be more favourable because the dependence of selected countries such as the USA or Brazil could not be alleviated in this way. By the energetic viewpoint, more than one alternative source would be needed in a more synergic strategy (solar energy, ethanol, biomasses consumption, etc.). In addition, the reduction of CF indexes should be considered in terms of kg per m^3 (or similar measure units). However, available data could not be always correlated with purely energetic considerations, because the source of energy used for all operations (including the removal of oxygen-consuming waste and materials) may depend on fossil fuels or renewable sources. As a simple consequence, should fuel consumption be needed, the energetic balance of ethanol production would be worsened. On the other hand, should renewable sources be used, the global impact would be lower but carbon dioxide levels could be relevant enough (de Oliveira et al. 2005).

Finally, CF is sometimes used with the aim of demonstrating that traditional farming activities are extremely favourable and environmentally sustainable on condition that transportation is reduced. On the contrary, the production of selected foods such as eggs in the USA is not favourable when speaking of CF indexes. In detail, a peculiar ratio may be demonstrated when comparing (a) calculated emissions from egg farms as kilograms per dozen eggs and (b) calculated emission derived from fuel consumption for delivery, in terms of litres per dozen eggs (Capper 2011). This ratio is not in favour of agricultural activities (2.32 times in comparison with transport activities). Actually, a similar ratio is demonstrable when comparing the same activities performed by Buyers' clubs (2.29:1) or grocery stores (2.33:1). Substantially, there are no evidences that reduced transportation (or a minor amount of 'food miles') also lowers the CF index (Edwards-Jones et al. 2008); on the contrary, the proportion between primary production and indirect activities (including delivery, etc.) is approximately 2.3:1.0 when speaking of CF and suspected environmental damages. These data (Fig. 2.1)—and other researches carried out in relation to agricultural activities—seem to confirm the main importance of basic primary production—both as agricultural activities and animal production—in terms of 'ecological' sustainability at least (Capper et al. 2009, b). In fact, delivery services may account only for 11% of the CF index, while the final transportation from food producers to retail services does not exceed 4% (Weber and Matthews 2008). Consequently, CF results may give an approximate estimation of the contribution to environmental damages in the agricultural sector only on condition that accessory services are considered out of the calculation. In this way, the CF contribution should be exactly evaluated: it has been reported that the maximum amount of CF values derived from primary production may reach 83%, in relation to the US household footprint (Weber and Matthews 2008).

2.4 Phosphorus and Agricultural Expansion. Current Problems and Possible Management

Phosphorus is generally considered when speaking of environmental damages with reference to aquatic eutrophication, as recently reported (Hooda et al. 2000; Levine and Schindler 1989). The problem of phosphorus is that contaminated water becomes progressively unsatisfactory for drinking and other purposes (Sharpley and Menzel 1987; Sharpley et al. 2000; USEPA 1990), including also agricultural activities, because of eutrophication effects (the reduction of bioavailable nutrients for non-algae organisms; the increased amount of carbon, nitrogen, and phosphorus; oxygen depletion; fish death; etc.).

In relation to blue waters (Sect. 2.2), phosphorus is generally originated from agricultural areas, farms, industries, precipitations, and municipal or industry-linked sewages (Hooda et al. 2000; Isermann 1990; Ryding et al. 1990). In absence of different anthropic activities and clear pollution sources, it should be assumed that phosphorus pollution—expressed as total phosphorus, molybdate reactive phosphorus, soluble reactive phosphorus, or dissolved phosphorus—is mainly originated from agricultural activities. The 'phosphorus footprint' (PF) can be expressed in terms of grams, or gigagrams per year (Wang et al. 2011), and evaluated in terms of efficient distribution in soils, plants, and animals (Günther et al. 2017). In fact, the CF can be defined as the amount of mined phosphate needed to meet food demand by a specified organism (Metson et al. 2016).

On the other hand, different expressions can be used when speaking of the simple P concentration in soils or waters, and research chemists are probably accustomed to use these expressions instead of PF estimations, when speaking of environmental problems (PF concerns mainly the efficient P availability of living microorganisms). For example, the expression in relation to soil contamination should be calculated as mg per kg (Hooda et al. 2000). As a result, phosphorus (P) concentrations in effluent waters (mg/l) depend on the initial P concentration in soils: it has been reported that critical levels for total and inorganic P are between 0.01 and 0.02 mg/l concerning eutrophication (Daniel et al. 1998). In general, two sources can be considered when considering elevated PF values:

(a) Natural abundance in soils
(b) Nature of the substrate, with reference to organic/sandy or fine-textured soils, or the possible presence of macro-pores
(c) Use of fertilisers
(d) Use of pesticides containing organophosphorus or synthetic pyrethroid compounds
(e) Use of P-rich manure, also named phosphate-rich organic manure.

Organic and sandy substrates tend to show higher leaching P losses (with consequent P release in drained water) because of the low amount of carbonates, clays, etc., if compared with fine-textured soils (carbonates and other compounds are responsible for P absorption). The possible presence of macro-pores can also give a higher P leaching with enhanced P amounts in waters (Hooda et al. 2000; Stamm et al. 1998). It has been reported that P losses could not exceed 5% of the total applied P on soils; however, because of possible eutrophication, similar quantities can be of interest. A possible solution against excessive P leaching could be the use of slow/controlled release fertilisers (Shaviv and Mikkelsen 1993; Thiex 2016), while other researchers suggest moving manure from surplus lands to deficient soils (Sharpley et al. 2000).

In addition, the use of P-based pesticides is cause for different problems, including fish death on a large scale (Hooda et al. 2000; Virtue and Clayton 1997). Anyway, it has been reported that both P amounts in water and PF show concerning increases in the most important industrialised areas of the world, including also developing areas (Metson et al. 2016). Consequently, it has been suggested that a good method to limit P excess should be the limitation of meat-intensive diets in favour of vegetable-based diets.

References

Amthor JS (1998) Perspective on the relative insignificance of increasing atmospheric CO_2 concentration to crop yield. Field Crops Res 58(2):109–127. https://doi.org/10.1016/s0378-4290(98)00089-6

Anonymous (2006a) UK schools carbon footprint scoping study for Sustainable Development Commission. Global Action Plan, Stockholm Environment Institute, and Eco-Logica Ltd. for the Sustainable Development Commission, London, Mar 2006. Available http://www.se-ed.co.uk/sites/default/files/resources/GAP-Final-Report.pdf. Accessed 10 Sept 2018

Anonymous (2006b) Counting consumption—CO_2 emissions, material flows and ecological footprint of the UK by region and devolved country. World Wildlife Fund (WWF)-UK, Godalming, 84 pp

Austin RB (1999) Yield of wheat in the United Kingdom: recent advances and prospects. Crop Sci 39(6):1604–1610. https://doi.org/10.2135/cropsci1999.3961604x

Barbier EB (2004) Explaining agricultural land expansion and deforestation in developing countries. Am J Agric Econ 86(5):1347–1353. https://doi.org/10.1111/j.0002-9092.2004.00688.x

Berg W, Brunsch R, Hellebrand HJ, Kern J (2006) Methodology for measuring gaseous emissions from agricultural buildings, manure, and soil surfaces. In: Proceedings of the workshop on agricultural air quality: state of the science, Bolger Conference Center, Potomac, U.S.A., vol 58, pp 233–241, 5–8 June 2006

Boardman B (2008) Carbon labelling: too complex or will it transform our buying? Signif 5(4):168–171. https://doi.org/10.1111/j.1740-9713.2008.00322.x

Boote KJ, Pickering NB, Allen LH Jr (1997) Plant modeling: advances and gaps in our capability to predict future crop growth and yield in response to global climate change. In: Allen LH Jr, Kirkham MB, Olszyk DM, Whitman CE (eds) Advances in carbon dioxide effects research. American Society of Agronomy, Madison, pp 179–228

BSI (2008) PAS2050: 2008, specification for the assessment of the life cycle greenhouse gas emissions of goods and services. British Standards Institution (BSI), London

Capper JL (2011) Replacing rose-tinted spectacles with a high-powered microscope: the historical versus modern carbon footprint of animal agriculture. Anim Front 1(1):26–32. https://doi.org/10.2527/af.2011-0009

Capper JL, Cady RA, Bauman DE (2009) The environmental impact of dairy production: 1944 compared with 2007. J Anim Sci 87(6):2160–2167. https://doi.org/10.2527/jas.2009-1781

Chapagain AK, Hoekstra AY, Savenije HHG, Gautam R (2006) The water footprint of cotton consumption: an assessment of the impact of worldwide consumption of cotton products on the water resources in the cotton producing countries. Ecol Econ 60(1):186–203. https://doi.org/10.1016/j.ecolecon.2005.11.027

Core Writing Team, Pachauri RK, Reisinger A (2007) Climate change 2007: synthesis report. Contribution of Working Groups I, II and III to the Fourth Assessment Report of the Intergovernmental Panel on Climate Change. Intergovernmental Panel on Climate Change (IPCC), Geneva, 104 pp. Available https://www.ipcc.ch/pdf/assessment-report/ar4/syr/ar4_syr_full_report.pdf. Accessed 10 Sept 2018

Daniel TC, Sharpley AN, Lemunyon JL (1998) Agricultural phosphorus and eutrophication: a symposium overview. J Environ Qual 27:251–257. https://doi.org/10.2134/jeq1998.00472425002700020002x

de Oliveira ME, Vaughan BE, Rykiel EJ Jr (2005) Ethanol as fuel: energy, carbon dioxide balances, and ecological footprint. BioSci 55(7):593–602. https://doi.org/10.1641/0006-3568(2005)055[0593:EAFECD]2.0.CO;2

DEFRA (2006) The environment in your pocket. Department for Environment, Food and Rural Affairs (DEFRA), London

Downing TE, Harrison PA, Butterfield RE, Lonsdale KG (1999) Climate change, climatic variability and agriculture in Europe: an integrated assessment. Research Report No. 21. Environmental Change Unit, University of Oxford, Oxford

Easterling WE, Mearns LO, Hays CJ, Marx D (2001) Comparison of agricultural impacts of climate change calculated from high and low resolution climate change scenarios. Part II. Accounting for adaptation and CO_2 direct effects. Clim Change 51, 2:173–197. https://doi.org/10.1023/a:1012267900745

Edwards-Jones G, Milà i Canals L, Hounsome N, Truninger M, Koerber G, Hounsome B, Cross P, York EH, Hospido A, Plassmann K, Harris IM, Edwards RT, Day GAS, Tomos AD, Cowell SJ, Jones DL (2008) Testing the assertion that 'local food is best': the challenges of an evidence-based approach. Trends Food Sci Technol 19:5:265–274. https://doi.org/10.1016/j.tifs.2008.01.008

Evans LT, Fischer RA (1999) Yield potential: its definition, measurement, and significance. Crop Sci 39(6):1544–1551. https://doi.org/10.2135/cropsci1999.3961544x

Ewert F, Rounsevell MDA, Reginster I, Metzger MJ, Leemans R (2005) Future scenarios of European agricultural land use: I. Estimating changes in crop productivity. Agric Ecosystems Environ 107:2–3:101–116. https://doi.org/10.1016/jagee.2004.12.003

FAOSTAT (2018) The Food and Agricultural Organization of the United Nations. Available http://faostat.fao.org/. Accessed 10 Sept 2008

Galli A, Wiedmann T, Ercin E, Knoblauch D, Ewing B, Giljum S (2012) Integrating ecological, carbon and water footprint into a "footprint family" of indicators: definition and role in tracking human pressure on the planet. Ecol Indic 16:100–112. https://doi.org/10.1016/j.ecolind.2011.06.017

Grigg OB (1993) The world food problem. Blackwell, Oxford

Grubb and Ellis (2007) Meeting the carbon challenge: the role of commercial real estate owners, users & managers. Grubb & Ellis Company, Chicago, Apr 2007

Günther J, Thevs N, Gusovius HJ, Sigmund I, Brückner T, Beckmann V, Abdusalik N (2017) Carbon and phosphorus footprint of the cotton production in Xinjiang, China, in comparison to an alternative fibre (Apocynum) from Central Asia. J Clean Prod 148(1):490–497. https://doi.org/10.1016/j.jclepro.2017.01.153

Hammerschlag R, Barbour W (2003) Life-cycle assessment and indirect emission reductions: issues associated with ownership and trading. Institute for Lifecycle Environmental Assessment (ILEA), Seattle, May 2003. Available http://www.ilea.org/downloads/LCAEmissReductions.pdf. Accessed 10 Sept 2018

Harrison PA, Butterfield RE (1996) Effects of climate change on Europe-wide winter wheat and sunflower productivity. Clim Res 7:225–241. https://doi.org/10.3354/cr007225

Hertwich EG, Peters GP (2009) Carbon footprint of nations: a global, trade-linked analysis. Environ Sci Technol 43(16):6414–6420. https://doi.org/10.1021/es803496a

Hoekstra A (2013) The water footprint of animal products: the meat crisis: developing more sustainable and ethical production and consumption. In: D'Silva J, Webster J (eds) The meat crisis. Routledge, London, pp 41–52

Hoekstra AY, Mekonnen MM (2012) The water footprint of humanity. Proc Natl Acad Sci USA 109(9):3232–3237. https://doi.org/10.1073/pnas.1109936109

Hoekstra AY, Chapagain AK, Aldaya MM, Mekonnen MM (2011) The water footprint assessment manual: Setting the global standard. Earthscan, London and Washington, DC. Available http://waterfootprint.org/media/downloads/TheWaterFootprintAssessmentManual_2.pdf. Accessed 10 Sept 2018

Hooda PS, Edwards AC, Anderson HA, Miller A (2000) A review of water quality concerns in livestock farming areas. Sci Total Environ 250(1–3):143–167. https://doi.org/10.1016/S0048-9697(00)00373-9

Isermann K (1990) Share of agriculture in nitrogen and phosphorus emissions into the surface waters of Western Europe against the background of their eutrophication. Fert Res 26(1–3):253–269. https://doi.org/10.1007/BF01048764

ISO (2006) ISO 14040:2006—environmental management—life cycle assessment—principles and framework. International Organization for Standardization, Geneva

Jamieson P, Porter J, Semenov M, Brooks R, Ewert F, Ritchie J (1999) Comments on "Testing winter wheat simulation models predictions against observed UK grain yields" by Landau et al. (1998). Agric Forest Meteorol 96, 1–3:157–161. https://doi.org/10.1016/s0168-1923(99)00060-x

Johnson DG (1999) The growth of demand will limit output growth for food over the next quarter century. Proc Natl Acad of Sci 96(11):5915–5920. https://doi.org/10.1073/pnas.96.11.5915

Kenny T, Gray NF (2008) Comparative performance of six carbon footprint models for use in Ireland. Environ Impact Assess Rev 29(1):1–6. https://doi.org/10.1016/j.eiar.2008.06.001

Kimball BA, Kobayashi K, Bindi M (2002) Responses of agricultural crops to free-air CO_2 enrichment. Adv Agron 77:293–368. https://doi.org/10.1016/s0065-2113(02)77017-x

Landau S, Mitchell RAC, Barnett V, Colls JJ, Craigon J, Moore KL, Payne RW (1998) Testing winter wheat simulation models' predictions against observed UK grain yields. Agric For Meteorol 89:85–99. https://doi.org/10.1016/s0168-1923(97)00069-5

Laurent A, Olsen SI, Hauschild MZ (2010) Carbon footprint as environmental performance indicator for the manufacturing industry. CIRP Ann 59(1):37–40. https://doi.org/10.1016/j.cirp.2010.03.008

Laurent A, Olsen SI, Hauschild MZ (2012) Limitations of carbon footprint as indicator of environmental sustainability. Environ Sci Technol 46(7):4100–4108. https://doi.org/10.1021/es204163f

Leach AM, Emery KA, Gephart J, Davis KF, Erisman JW, Leip A, Pace ML, D'Odorico P, Carr J, Cattell Noll L, Castner E, Galloway JN (2016) Environmental impact food labels combining carbon, nitrogen, and water footprints. Food Policy 61:213–223. https://doi.org/10.1016/j.foodpol.2016.03.006

Lenzen M (2008) Double-counting in life cycle calculations. J Ind Ecol 12(4):583–599. https://doi.org/10.1111/j.1530-9290.2008.00067.x

Lenzen M, Murray J, Sack F, Wiedmann T (2007) Shared producer and consumer responsibility—theory and practice. Ecol Econ 61(1):27–42. https://doi.org/10.1016/j.ecolecon.2006.05.018

Levine SL, Schindler DW (1989) Phosphorus, nitrogen and carbon dynamics of experimental lake 303 during recovery from eutrophication. Can J Fish Aquat Sci 46(1):2–10. https://doi.org/10.1139/f89-001

Long SP (1991) Modification of the response of photosynthesis productivity to rising temperature by atmospheric CO_2 concentration: has its importance been underestimated? Plant Cell Environ 14(8):729–739. https://doi.org/10.1111/j.1365-3040.1991.tb01439.x

Matson PA, Parton WJ, Power AG, Swift MJ (1997) Agricultural intensification and ecosystem properties. Science 25:277(5325):504–509. https://doi.org/10.1126/science.277.5325.504

Mekonnen MM, Hoekstra AY (2011) A global assessment of the green, blue and grey water footprint of crops and crop products. Hydrol Earth Syst Sci 15:1577–1600. https://doi.org/10.5194/hess-15-1577-2011

Metson GS, Cordell D, Ridoutt B (2016) Potential impact of dietary choices on phosphorus recycling and global phosphorus footprints: the case of the average Australian city. Front Nutr 3:35. https://doi.org/10.3389/fnut.2016.00035

Meyer WB, Turner BL (1992) Human population growth and global land-use/cover change. Ann Rev Ecol Syst 23(1):39–61. https://doi.org/10.1146/annurev.es.23.110192.000351

Morison JIL, Lawlor DW (1999) Interactions between increasing CO_2 concentration and temperature on plant growth. Plant Cell Environ 22(6):659–682. https://doi.org/10.1046/j.1365-3040.1999.00443.x

Naylor RL (1996) Energy and resource constraints on intensive agricultural production. Ann Rev Energy Environ 21(1):99–123. https://doi.org/10.1146/annurev.energy.21.1.99

Oerke EC, Dehne HW (1997) Global crop production and the efficacy of crop protection-current situation and future trends. Eur J Plant Pathol 103(3):203–215. https://doi.org/10.1023/A:1008602111248

Padgett JP, Steinemann AC, Clarke JH, Vandenbergh MP (2008) A comparison of carbon calculators. Environ Impact Assess Rev 28:106–115. https://doi.org/10.1016/j.eiar.2007.08.001

Pandey D, Agrawal M, Pandey JS (2011) Carbon footprint: current methods of estimation. Environ Environ Monit Assess 178(1–4):135–160. https://doi.org/10.1007/s10661-010-1678-y

Parry M, Rosenzweig C, Iglesias A, Livermore M, Fischer G (2004) Effects of climate change on global food production under SRES emissions and socio-economic scenarios. Glob Environ Change 14(1):53–67. https://doi.org/10.1016/j.gloenvcha.2003.10.008

Peters GP (2010) Carbon footprints and embodied carbon at multiple scales. Curr Opin Environ Sustain 2(4):245–250. https://doi.org/10.1016/j.cosust.2010.05.004

Pfister S, Koehler A, Hellweg S (2009) Assessing the environmental impacts of freshwater consumption in LCA. Environ Sci Technol 43(11):4098–4104. https://doi.org/10.1021/es802423e

POST (2006) Carbon footprint of electricity generation. POST number 268, Oct 2006, pp 1–4. Parliamentary Office of Science and Technology (POST), London. Available https://www.parliament.uk/documents/post/postpn268.pdf. Accessed 10 Sep 2018

Rees WE (1996) Revisiting carrying capacity: area-based indicators of sustainability. Popul Environ 17(3):195–215. https://doi.org/10.1007/bf02208489

Reynolds MP, Rajaram S, Sayre KD (1999) Physiological and genetic changes of irrigated wheat in the post-green revolution period and approaches for meeting projected global demand. Crop Sci 39(6):1611–1621. https://doi.org/10.2135/cropsci1999.3961611x

Ridoutt BR, Sanguansri P, Harper GS (2011) Comparing carbon and water footprints for beef cattle production in Southern Australia. Sustain 3:2443–2455. https://doi.org/10.3390/su3122443

Rosegrant MW, Paisner MS, Meijer S, Witcover J (2001) Global food projections to 2020. Emerging trends and alternative futures. International Food Policy Research Institute (IFPRI), Washington, D.C

Rosenzweig C, Parry ML (1994) Potential impact of climate change on world food supply. Nature 367(6459):133–138. https://doi.org/10.1038/367133a0

Rounsevell MDA, Annetts JE, Audsley E, Mayr T, Reginster I (2003) Modelling the spatial distribution of agricultural land use at the regional scale. Agric Ecosyst Environ 95(2–3):465–479. https://doi.org/10.1016/s0167-8809(02)00217-7

Ryding SO, Enell M, Wennberg L (1990) Swedish agricultural nonpoint source pollution: a summary of research and findings. Lake Reserv Manag 6(2):207–217. https://doi.org/10.1080/07438149009354711

Schiermeier Q (2006) Climate credits. Nature 444(7122):976–977. https://doi.org/10.1038/444976a

Segal R, MacMillan T (2009) Water labels on food. Issues and recommendations. Food Ethics Council, Brighton. Available https://www.foodethicscouncil.org/uploads/publications/waterlabels_0.pdf. Accessed 10 Sept 2018

Sharpley AN, Menzel RG (1987) The impact of soil fertiliser phosphorus on the environment. Adv Agron 41:297–324. https://doi.org/10.1016/S0065-2113(08)60807-X

Sharpley A, Foy B, Withers P (2000) Practical and innovative measures for the control of agricultural phosphorus losses to water: an overview. J Environ Qual 29(1):1–9. https://doi.org/10.2134/jeq2000.00472425002900010001x

Shaviv A, Mikkelsen RL (1993) Controlled-release fertilizers to increase efficiency of nutrient use and minimize environmental degradation—a review. Fert Res 35(1–2):1–12. https://doi.org/10.1007/BF00750215

Stamm CH, Flühler H, Gächter R, Leuenberger J, Wunderli H (1998) Preferential transport of phosphorus in drained grassland soils. J Environ Qual 27(3):515–522. https://doi.org/10.2134/jeq1998.00472425002700030006x

Steen-Olsen K, Weinzettel J, Cranston G, Ercin AE, Hertwich EG (2012) Carbon, land, and water footprint accounts for the European Union: consumption, production, and displacements through international trade. Environ Sci Technol 46(20):10883–10891. https://doi.org/10.1021/es301949t

Thiex N (2016) Determination of nitrogen, phosphorus, and potassium release rates of slow-and controlled-release fertilizers: single-laboratory validation, first action 2015.15. J AOAC Int 99, 2:353–359. https://doi.org/10.5740/jaoacint.15-0294

Trucost (2006) Carbon counts: the Trucost carbon footprint ranking of UK Investment Funds. Trucost Ltd, London

Tubiello FN, Ewert F (2002) Simulating the effects of elevated CO_2 on crops: approaches and applications for climate change. Eur J Agron 18(1–2):57–74. https://doi.org/10.1016/s1161-0301(02)00097-7

UN (1998) Kyoto protocol to the United Nations framework convention on climate change (UNFCCC). United Nations (UN), New York, 21 pp. Available https://unfccc.int/sites/default/files/kpeng.pdf. Accessed 10 Sept 2018

UN News (2007) Climate change 'defining issue of our era' says Ban Kimoon, hailing G8 action. 2007. United Nations (UN), New York. https://news.un.org/en/story/2007/06/221622-climate-change-defining-issue-our-era-says-ban-ki-moon-hailing-g8-action. Accessed 07 Sept 2018

USCCTP (2005) Technology options for near and long term future. United States Climate Change Technology Program (USCCTP), Washington, DC

USEPA (1990) National water quality inventory 1988. United States Environmental Protection Agency (USEPA) Report to Congress. Office of Water, US Govt. Print Office, Washington, DC

Velasco E, Pressley S, Allwine E, Westberg H, Lamb B (2005) Measurements of CO_2 fluxes from the Mexico City urban landscape. Atmos Environ 39(38):7433–7446. https://doi.org/10.1016/j.atmosenv.2005.08.038

Virtue WA, Clayton JW (1997) Sheep dip chemicals and water pollution. Sci Total Environ 194–195:207–217. https://doi.org/10.1016/S0048-9697(96)05365-X

Wang F, Sims JT, Ma L, Ma W, Dou Z, Zhang F (2011) The phosphorus footprint of China's food chain: implications for food security, natural resource management, and environmental quality. J Environ Qual 40(4):1081–1089. https://doi.org/10.2134/jeq2010.0444

Weber CL, Matthews HS (2008) Food-miles and the relative climate impacts of food choices in the United States. Environ Sci Technol 42(10):3508–3513. https://doi.org/10.1021/es702969f

Wiedmann T, Minx J (2008) A definition of 'carbon footprint'. In: Pertsova CC (ed) Ecological economics research trends (Chap. 1), Nova Science Publishers, Hauppauge, pp 1–11

Chapter 3
Bee Viruses and the Related Impact on Food Crops Worldwide

Abstract This chapter evaluates the impact of bee viruses on food crops and the predictable effect on food production. The production of crops worldwide can depend on many factors including temperature, moisture, salinity, pH of the soil, and the attack by biological agents such as aphids or viruses. With reference to the last problem, honeybees can act as target organisms. Because of the demonstrated correlation between the activity of honeybees, also named pollination, and the yield of certain crops (fruits, vegetables, and nuts), it can be also inferred that negative factors lowering pollination can cause the rise of food prices for vegetable raw materials. Therefore, new preventive approaches such as the surveillance of bees should be considered for prediction purposes, especially when speaking of factors influencing negatively pollination: the attack of bee viruses. The observed diminution of honeybees because of high mortality rates—the colony collapse disorder—has requested new studies concerning the use of pesticides, antibiotic therapies, and possible genetic selection strategies.

Keywords Beekeeper · Bee virus · Colony collapse disorder · *Varroa* · Honeybee · Neonicotinoid · Pollination

Abbreviations

ABPV   Acute bee paralysis virus
CCD   Colony collapse disorder
IAPV   Israeli acute paralysis virus
KBV   Kashmir bee virus

© The Author(s), under exclusive license to Springer Nature Switzerland AG 2019   31
S. D. Sharma et al., *Raw Material Scarcity and Overproduction in the Food Industry*,
Chemistry of Foods, https://doi.org/10.1007/978-3-030-14651-1_3

3.1   The Modern Crop Production and the Influence of HoneyBees. An Overview

In general, the production of crops worldwide can depend on many factors including (Ahmad et al. 2014; Barney and Bedford 2008; Kozai and Niu 2016):

(1) Temperature
(2) Moisture
(3) Salinity
(4) Drought
(5) Presence of pesticides
(6) pH of the soil
(7) Abundance (or deficiency, on the other hand) of bioavailable nitrogen, phosphorus, potassium, and sulphur
(8) Presence of heavy metals
(9) Use of traditional (open-field) farming, or
(10) Use of new farming procedures such as indoor vertical farming with artificial lighting
(11) Attack by biological agents such as aphids or viruses.

The management of correlated activities and farming systems has to be considered; similar factors could be used for the creation of dedicated models (Nemecek and Gaillard 2010).

One of the above-mentioned factors of interest is the possibility of attack by agents such as viruses. In general, virus diseases concern certain peculiar crops such as potatoes (Finch et al. 2014). However, some peculiar crop productions are directly influenced by the activity of honeybees. In detail, the 'pollination' is an important co-factor (Rucker et al. 2011) influencing positively and enhancing the production of many fruits, vegetables, and nuts such as apples, almonds, blueberries, pears, pumpkins, etc. (Holt 2014). The positive influence of pollination is observed worldwide with an estimated 300%—augment of pollination-dependent crop production since 1961 (Aizen and Harder 2009). Consequently, it may be inferred that a demonstrable correlation exists between the global crop production (this amount depends on the pollination by bees) and rising food prices of raw materials. Consequently, new preventive approaches such as the surveillance of bees should be considered for prediction purposes, especially when speaking of factors influencing negatively pollination: the attack of bee viruses.

3.2   The Pollination

In general, 'pollination' means the process of pollen grains transfer from the male anther of a flower to the female stigma (pistil) of another flower (Anonymous 2018), with the aim of making seeds. For this reason, flowers of the same species have to be involved.

Actually, 'pollinator agents' should be recognised as living vectors able to transfer pollen grains from one to another flower (Anonymous 2018): wind, water, bats, birds, butterflies, insects, etc. On the other side, pollination might rely on inanimate (abiotic) vectors—wind or water is good example—with good efficacy when speaking of fertilisation yields (Faegri and Van der Pijl 1979; McGregor 1976).

Anyway, the pollination may concern reproductive systems of two different flowers (cross-pollination) or of the same flower (self-pollination). In this situation, the fertilisation of the female stigma occurs in the same flower.

By the viewpoint of flower growers, the pollination can be an extremely powerful—and historically recognised—system when speaking of agricultural practices (Faegri and Van der Pijl 1979; Frankel and Esra Galun 1977). This statement is especially true when speaking of pollination performed by honeybees. In detail, it should be considered that honeybees pollinate 90% of flowering plants, while the fertilisation of the resting part of plants is ascribed to abiotic agents: wind and water (Rucker et al. 2011). It has been reported that bee-keeping was known at least 5000 years ago (DeGrandi-Hoffman 2003; Rucker et al. 2011). At present, pollination by honeybees is a common practice both in European countries and in the North America. It has to be considered also that hybridisation between different species of the same plant (usually, tree nuts and fruits) is possible by means of cross-pollination, provided that different plant species are located in different places. This farming strategy implies that different plants are placed in adjacent rows (Rucker et al. 2011). According to recent studies (Seeley 1995), it may be affirmed that honeybees tend to cover a circular area with the colony as centre and the radius of 6 km, even if the maximum distance from colonies and bee forages could reach 10.9 km (Fig. 3.1).

This situation has favoured and enhanced a peculiar 'pollination' sector performed by mobile bee-keepers, in North America at least. It should be noted that mobile bee-keepers usually follow certain migratory routes followed by bees, and these routes are typical of geographical areas, in North America at least (Burgett et al. 2009, 2010; Ferrier et al. 2018). For these reasons, it may be assumed that a direct correlation exists between the activity of honeybees—and related bee-keepers—on the one side and the production of selected crops on the other side. However, it could be supposed that the higher the number of active honeybees, the higher the result in terms of crop yields. On the contrary, recent studies and the examination of production results, year per year, demonstrate clearly that a certain overestimation is possible and maybe predictable. Causes are not always dependent on honeybees, but their activity is able to influence directly obtained results.

The reduction—whether unexpected or not—of crop productions can be correlated with the possible diminution of living honeybees. This phenomenon, originally observed in 2007, was called 'Colony Collapse Disorder' (CCD) in the United States of America (Pasciak 2013; Rucker et al. 2011; vanEngelsdorp et al. 2007, 2008, 2009, 2010; Williams et al. 2010). It has been reported that the total amount of crop pro-

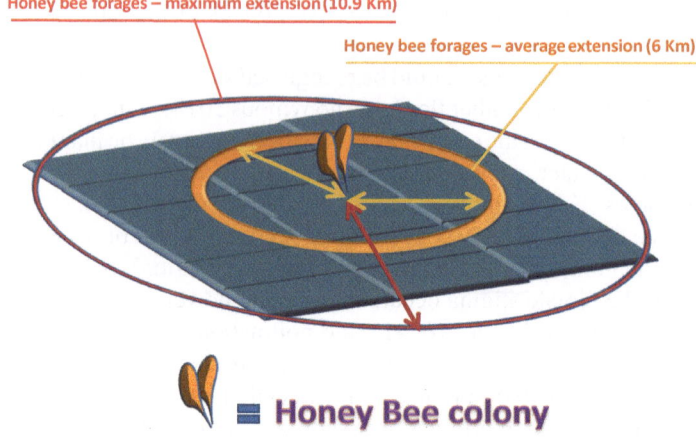

Fig. 3.1 By the viewpoint of flower growers, the pollination can be an extremely powerful system when speaking of agricultural practices. This statement is especially true when speaking of pollination performed by honey bees (they pollinate 90% of flowering plants, while the fertilisation of the resting part of plants is ascribed to abiotic agents: wind and water). Moreover, hybridisation between different species of the same plant is possible by means of cross-pollination, provided that different plant species are located in different places. According to recent studies, it may be affirmed that honey bees tend to cover a circular area with the colony as centre and the radius of 6 km, even if the maximum distance from colonies and bee forages could reach 10.9 km (Seeley 1995)

duction losses, both direct and indirect results, could reach potentially $90 billion in relation to the area of United States of America (Rucker et al. 2011; Stipp 2007). In summary, CCD can have a direct impact on pollination and consequently on crop production yields.

3.3   HoneyBees and Related Enemies

In general, CCD is a phenomenon with several possible causes, both natural and anthropic factors. With concern to natural and 'historical' causes, the following factors should be considered (Botías et al. 2013; Ferrier et al. 2018; Holt 2014; Rucker et al. 2011):

(a) Bee parasites (causing related diseases) such as the fungus *Nosema ceranae* and *N. apis*.
(b) Bacterial infections, such as the American foulbrood (*Melissococcus plutonius*) and the European foulbrood (*Paenibacillus larvae*).
(c) Bee viruses, such as Israeli acute paralysis virus and *Iridoviridae* viruses. Deformed wing virus and Kashmir bee virus should be also considered as an important portion of synergic attackers because of their role in action with certain parasites (Shen et al. 2005).

(d) Other undefined bee diseases.

The first of these points concerns the spread of certain and extensively studied parasites, including at least *Acarapis woodi*, *Tropilaelaps clareae*, *Varroa destructor*, and *Varroa jacobsoni* (Bowen-Walker et al. 1999; Rucker et al. 2011; Steinhauer et al. 2014). The last of these parasites has been extensively studied in correlation with the presence of *Nosema* fungi and/or certain pesticides, as explained below.

On the other hand, anthropic reasons can be investigated, including:

(1) The use of certain nicotine-derived pesticides such as neonicotinoids (Bromenshenk et al. 2010; Holt 2014). Interestingly, the impact of these insecticides has been studied in relation to the crop production of different species, including canola, corn, cotton, soy, cotton, sorghum, rice, and barley (Holt 2014). The aim of these studies is generally to correlate *Nosema* diseases and related effects on yield productions when speaking of the concomitant use of peculiar insecticides. In addition, one of the above-mentioned crops—canola—is reported to be good forage for honeybees, with interesting effects when speaking of anti-*Nosema* preventive strategies (Holt 2014). In particular, three neonicotinoid representative molecules—clothianidin, imidacloprid, and thiamethoxam—were initially signalled in Europe as a powerful cause for bee deaths, and their presence on treated corn is a notable concern (European Commission 2013; Krupke et al. 2012). The negative effect of these pesticides has been repeatedly correlated with *Varroa* action.

(2) Malnutrition, when speaking of bee-keeping companies, and contaminated forages. The problem of *Nosema*-contaminated canola could be discussed (Biesmeijer 2006).

(3) Stress causes (probably caused by frequent travels for pollination, concerning the activity of bee-keepers).

(4) Insufficient plant diversity (Holt 2014; Kremen et al. 2002; Naug 2009).

3.4 Bee Viruses and Parasites

Nosema cerana was originally a virus concerning only the Asian species *Apis cerana* in East Asia and South Asia regions only: at present, its diffusion is demonstrated in Europe and North America. It is believed that *N. cerana* is able to use *Apis mellifera* as a host, although *N. cerana* pathogenicity against this common bee has to be evaluated.

At present, CCD is supposed to be caused by a synergic action by one virus—an *Iridoviridae* virus—on the one side, and the microsporidian *N. ceranae* on the other; however, the mechanism of this synergic—and powerful—action between the virus and the fungus have to be investigated (Bromenshenk et al. 2010; Johnson 2010).

Consequently, the only action of bee viruses is not demonstrated as the main cause of bee mortality. On the contrary, it may be inferred that selected viruses have a lethal action when in connection with *Varroa* or *Nosema* agents (natural factors), or neonicotinoids (and precipitation) as anthropic agents (Allen and Ball 1996; Holt 2014). With reference to *Varroa*, this parasite has been recognised as a vector for these viruses. Consequently, anti-*Varroa* fight is the best strategy against bee virus attacks (Ferrier et al. 2018; Pasciak 2013).

The list of more dangerous bee viruses is not long, containing the following names, mainly reported after 1980 (Ferrier et al. 2018; Pasciak 2013):

(a) Deformed wing virus (it is considered one of the main viruses when speaking of bee mortality. Anyway, it is considered the most prevalent virus at present)
(b) Sacbrood virus (moderate danger)
(c) Black queen cell virus (moderate danger)
(d) Chronic bee paralysis virus(moderate danger)
(e) Acute bee paralysis virus (ABPV, moderate danger)
(f) Israeli acute paralysis virus (IAPV, moderate danger, although it has been reported to be correlated positively with CCD episodes)
(g) Kashmir bee virus (KBV, moderate danger).

Actually, 18 different viruses have been reported in this ambit (Allen and Ball 1996). Interestingly, all mentioned viruses with the only exception of deformed wing virus are defined to be synergistic viruses (Benjeddou et al. 2001; Evans 2001; Hung et al. 1996a, b; Leat et al. 2000).

From the economic viewpoint, it is difficult to understand the real weight of CCD and bee diseases (caused by bee viruses in particular). Some economic study and related simulation revealed that fees concerning almond pollination could rise up to 16.7% if there is a CCD in progress. Consequently, it might be inferred that the estimated increase of prices on almonds in cans could reach 0.04% as derived exclusively as the effect of CCD for the peculiar sector of almonds. This estimation may be performed in many situations. However, it should be noted that (Burgett et al. 2009; Rucker et al. 2011):

(a) The impact of bee viruses as the main concern and cause of bee population declines should be investigated carefully and in relation to different geographical areas (the above-discussed example concerned only the North American regions). At present, the attack of bee viruses has been reported and extensively studied in countries including the United States of America, Denmark, Norway, Sweden, Hungary, France, Austria, Uruguay, Australia, United Kingdom, Japan, Germany, Greece, and China (Ai et al. 2012; Anderson and Gibbs 1988; Antúnez et al. 2006; Bacandritsos et al. 2010; Carreck et al. 2010; Ferrier et al. 2018; Forgách et al. 2008; Gensersch et al. 2010; Francis and Kryger 2012; Nordström et al. 1999; Ratnieks and Carreck 2010). One of the most known causes is generally reported to be described as 'globalised trade and travel' (Smith et al. 2013).

(b) CCD causes may be synergistic, and the result of non-virus attacks could be different from expected estimations.

(c) The mortality of bees as estimated by bee-keepers is generally 14%. Should external cause derived from bee viruses and/or fungi be expected, the decline of bee populations is expected to reach 30%. This damage can be translated in economic losses both for bee-keepers (lost bees have to be replaced) and the final customer renting bee-keeping services.

(d) Anyway, the economic impact of CCD and/or similar diseases on the world of food production—in terms of increase in shelf prices on the one side, and reduced crop yields on the other side—may be expected to be very different from these simulated estimations. As a simple example, the diesel price for bee-keepers may have a non-secondary weight. More research is needed in this difficult sector. At present, it has been reported that the known CCD effect has only a modest effect on food prices (Holt 2014; Rucker et al. 2011).

(e) The price of honey can be an interesting indicator. It has been reported that the North American CCD causes an approximate 100%—increase when speaking of honey prices after eight years of demonstrated CCD (Holt 2014).

(f) Finally, economy is different when speaking of macro-economic areas. The same thing has to be considered when speaking of relations between bee diseases and pollination damages in different world regions.

3.5 The Economic Impact of Bee Viruses and Other Agents. Possible Strategies

Different agents require different strategies. It has been reported that (Ferrier et al. 2018):

(1) Bacterial infections attacking bees are treated prophylactically with antibiotic compounds (application time: five days). In general, thymol and terramycin appear good enough in this ambit. However, colonies attacked by the American foulbrood are destroyed because of the high re-infection potential.

(2) *Varroa* and other parasites are generally treated with coumaphos, amitraz, and/or thymol. The use of oxalic acid, menthol crystals, vegetable oils, or sugar is signalled because viruses are better and preventively contrasted when preventing *Varroa*.

(3) Fungicides are used against fungi (*N. ceranae* and *N. apis*).

Other strategies may concern the fight against insect parasites such as *Aethina tumida* (the small hive beetle) or the application of genetic methods with the aim of selecting 'survivor' stocks against multiple menaces. Probably, genetic selection is the future when speaking of bee mortality and contrasting methods because of the possibility to fight different agents at the same time with a few survivor stocks only.

On the other side, there is no confirmation that genetically engineered crops—for forage purposes, in relation to bees and related nutrition—may solve the problem of concomitant contamination by fungi, parasites, and the presence of viruses transported by *Varroa* mites and other similar agents (Ferrier et al. 2018).

References

Ahmad P, Jamsheed S, Hameed A, Rasool S, Sharma I, Azooz M, Hasanuzzaman M (2014) Drought stress induced oxidative damage and antioxidants in plants. In: Ahmad P (ed) Oxidative damage to plants. Academic Press, San Diego, pp 345–367. https://doi.org/10.1016/c2013-0-06923-x

Ai H, Yan X, Han R (2012) Occurrence and prevalence of seven bee viruses in *Apis mellifera* and *Apis cerana* apiaries in China. J Invertebr Pathol 109(1):160–164. https://doi.org/10.1016/j.jip.2011.10.006

Aizen MA, Harder LD (2009) The global stock of domesticated honey bees is growing slower than agricultural demand for pollination. Curr Biol 19(11):915–918. https://doi.org/10.1016/j.cub.2009.03.071

Allen M, Ball B (1996) The incidence and world distribution of honey bee viruses. Bee World 77(3):141–162. https://doi.org/10.1080/0005772X.1996.11099306

Anderson DL, Gibbs AJ (1988) Inapparent virus infections and their interactions in pupae of the honey bee (*Apis mellifera* Linnaeus) in Australia. J Gen Virol 69(7):1617–1625. https://doi.org/10.1099/0022-1317-69-7-1617

Anonymous (2018) What is pollination? United States Department of Agriculture, Forest Service, Washington, DC. Available https://www.fs.fed.us/wildflowers/pollinators/What_is_Pollination/index.shtml. Accessed 11 Sept 2018

Antúnez K, D'Alessandro B, Corbella E, Ramallo G, Zunino P (2006) Honeybee viruses in Uruguay. J Invertebr Pathol 93(1):67–70. https://doi.org/10.1016/j.jip.2006.05.009

Bacandritsos N, Granato A, Budge G, Papanastasiou I, Roinioti E, Caldon M, Falcaro C, Gallina A, Mutinelli F (2010) Sudden deaths and colony population decline in Greek honey bee colonies. J Invertebr Pathol 105(3):335–340. https://doi.org/10.1016/j.jip.2010.08.004

Barney D, Bedford L (2008) Raw material selection: fruit, vegetables and cereals. In: Brown M (ed) Chilled foods, 3rd edn. Woodhead Publishing Ltd., Cambridge

Benjeddou M, Leat N, Allsopp M, Davison S (2001) Detection of acute bee paralysis virus and black queen cell virus from honeybees by reverse transcriptase PCR. Appl Environ Microbiol 67(5):2384–2387. https://doi.org/10.1128/aem.67.5.2384-2387.2001

Biesmeijer JC (2006) Parallel declines in pollinators and insect-pollinated plants in Britain and the Netherlands. Science 313(5785):351–354. https://doi.org/10.1126/science.1127863

Botías C, Martín-Hernández R, Barrios L, Meana A, Higes M (2013) *Nosema* spp. infection and its negative effects on honey bees (*Apis mellifera iberiensis*) at the colony level. Vet Res 44(1):25. https://doi.org/10.1186/1297-9716-44-25

Bowen-Walker P, Martin S, Gunn A (1999) The transmission of deformed wing virus between honeybees (*Apis mellifera* L.) by the ectoparasitic mite *Varroa jacobsoni* oud. J Invertebr Pathol 73(1):101–106. https://doi.org/10.1006/jipa.1998.4807

Bromenshenk JJ, Henderson CB, Wick CH, Stanford MF, Zulich AW, Jabbour RE, Deshpande SV, McCubbin PE, Seccomb RA, Welch PM, Williams T, Firth DR, Skowronski E, Lehmann MM, Bilimoria SL, Gress J, Wanner KW, Cramer RA (2010) Iridovirus and microsporidian linked to honey bee colony decline. PLoS ONE 5(10):e13181. https://doi.org/10.1371/journal.pone.0013181

Burgett M, Rucker RR, Thurman W (2009) Honey bee colony mortality in the Pacific northwest (USA). Am Bee J 149(6):573–575

Burgett M, Daberkow S, Rucker RR, Thurman W (2010) U.S. pollination markets: recent changes and historical perspective. Am Bee J 150(1):35–41

Carreck NL, Ball BV, Martin SJ (2010) Honey bee colony collapse and changes in viral prevalence associated with *Varroa destructor*. J Apic Res 49(1):93–94. https://doi.org/10.3896/ibra.1.49.1.13

DeGrandi-Hoffman G (2003) Honey bees in U.S. agriculture: past, present, and future, chap 1. In: Strickler K, Cane JH (eds) For nonnative crops, whence pollinators of the future? Thomas Say Publications in Entomology, Lanham, pp 204

European Commission (2013) Bee health: EU-wide restrictions on pesticide use to enter into force on 1 December. European Commission—press release, May 24. European Commission, Brussels. Available http://europa.eu/rapid/press-release_IP-13-457_en.htm?locale=en. Accessed 11 Sept 2018

Evans JD (2001) Genetic evidence for coinfection of honey bees by acute bee paralysis and Kashmir bee viruses. J Invertebr Pathol 78(4):189–193. https://doi.org/10.1006/jipa.2001.5066

Faegri K, Van der Pijl L (1979) Applied pollination technology. In: The principles of pollination ecology, 3rd edn. Pergamon Press, Oxford, New York, Toronto, Sidney, Paris, and Frankfurt

Ferrier PM, Rucker RR, Thurman WN, Burgett M (2018) Economic effects and responses to changes in honey bee health. Economic research report number 246, Mar 2018, pp 54. United States Department of Agriculture, Economic Research Service, Washington, DC. Available https://www.ers.usda.gov/webdocs/publications/88117/err-246.pdf?v=43186. Accessed 11 Sept 2018

Finch HJS, Samuel AM, Lane GPF (2014) Plant breeding and seed production. In: Lockhart & Wiseman's crop husbandry including Grassland. Woodhead Publishing Ltd., Cambridge, pp 263–283. https://doi.org/10.1533/9781782423928.2.263

Forgách P, Bakonyi T, Tapaszti Z, Nowotny N, Rusvai M (2008) Prevalence of pathogenic bee viruses in Hungarian apiaries: situation before joining the European Union. J Invertebr Pathol 98(2):235–238. https://doi.org/10.1016/j.jip.2007.11.002

Francis R, Kryger P (2012) Single assay detection of acute bee paralysis virus, Kashmir bee virus and Israeli acute paralysis virus. J Apic Sci 56(1):137–146. https://doi.org/10.2478/v10289-012-0014-x

Frankel R, Esra Galun E (1977) Pollination mechanisms, reproduction and plant breeding. Springer International Publishing, Cham

Genersch E, von der Ohe W, Kaatz H, Schroeder A, Otten C, Büchler R, Berg S, Ritter W, Mühlen W, Gisder S, Meixner M, Liebig G, Rosenkranz P (2010) The German bee monitoring project: a long term study to understand periodically high winter losses of honey bee colonies. Apidologie 41(3):332–352. https://doi.org/10.1051/apido/2010014

Holt JM (2014) The effects of environmental factors on honey bee morbidity. Dissertation, University of Illinois, Urbana. Available https://www.ideals.illinois.edu/bitstream/handle/2142/50359/Jai_Holt.pdf?sequence=1&isAllowed=y. Accessed 11 Sept 2018

Hung ACF, Ball BV, Adams JR, Shimanuki H, Knox DA (1996a) A scientific note on the detection of American strains of acute paralysis virus and Kashmir bee virus in dead bees in one US honey bee (*Apis mellifera* L.) colony. Apidologie 27(1):55–56. https://doi.org/10.1051/apido:19960107

Hung ACE, Shimanuki H, Knox DA (1996b) Inapparent infection of acute paralysis virus and Kashmir bee virus in the U.S. honey bees. Am Bee J 136:874–876

Johnson K (2010) Scientists and soldiers solve a bee mystery. New York Times, 6 Oct 2010. http://www.nytimes.com/2010/10/07/science/07bees.html. Accessed 11 Sept 2018

Kozai T, Niu G (2016) Introduction. In: Kozai T, Niu G, Takagaki M (eds) Plant factory—an indoor vertical farming system for efficient quality food production. Academic Press, London, San Diego, Waltham, and Oxford, pp 3–5. https://doi.org/10.1016/b978-0-12-801775-3.00001-9

Kremen C, Williams NM, Thorp RW (2002) Crop pollination from native bees at risk from agricultural intensification. Proc Natl Acad Sci 99(26):16812–16816. https://doi.org/10.1073/pnas.262413599

Krupke CH, Hunt GJ, Eitzer BD, Andino G, Given K (2012) Multiple routes of pesticide exposure for honey bees living near agricultural fields. PLoS ONE 7(1):e29268. https://doi.org/10.1371/journal.pone.0029268

Leat N, Govan V, Davison S, Ball B (2000) Analysis of the complete genome sequence of black queen-cell virus, a picorna-like virus of honey bees. J Gen Virol 81(8):2111–2119. https://doi.org/10.1099/0022-1317-81-8-2111

McGregor SE (1976) Insect pollination of cultivated crop plants, vol 496. United States Department of Agriculture, Agricultural Research Service, Washington, DC, p 411

Naug D (2009) Nutritional stress due to habitat loss may explain recent honeybee colony collapses. Biolog Conserv 142(10):2369–2372. https://doi.org/10.1016/j.biocon.2009.04.007

Nemecek T, Gaillard G (2010) Challenges in assessing the environmental impacts of crop production and horticulture. In: Environmental assessment and management in the food industry—life cycle assessment and related approaches. Woodhead Publishing Ltd., Cambridge, pp 98–116. https://doi.org/10.1533/9780857090225.2.98

Nordström S, Fries I, Aarhus A, Hansen H, Korpela S (1999) Virus infections in Nordic honey bee colonies with no, low or severe *Varroa jacobsoni* infestations. Apidologie 30(6):475–484. https://doi.org/10.1051/apido:19990602

Pasciak JL (2013) Crops, pesticides, and honey bee disease. Dissertation, University of Illinois, Urbana

Ratnieks FLW, Carreck NL (2010) Clarity on honey bee collapse? Science 327(5962):152–153. https://doi.org/10.1126/science.1185563

Rucker RR, Thurman WN, Burgett M (2011) Colony collapse: the economic consequences of bee disease. Montana State University, Department of Agricultural Economics and Economics

Seeley TD (1995) The wisdom of the hive: the social physiology of honey bee colonies. Harvard University Press, Cambridge

Shen M, Yang X, Cox-Foster D, Cui L (2005) The role of *Varroa* mites in infections of Kashmir bee virus (KBV) and deformed wing virus (DWV) in honey bees. Virology 342(1):141–149. https://doi.org/10.1016/j.virol.2005.07.012

Smith KM, Loh EH, Rostal MK, Zambrana-Torrelio CM, Mendiola L, Daszak P (2013) Pathogens, pests, and economics: drivers of honey bee colony declines and losses. EcoHealth 10(4):434–445. https://doi.org/10.1007/s10393-013-0870-2

Steinhauer NA, Rennich K, Wilson ME, Caron DM, Lengerich EJ, Pettis JS, Rose R, Skinner JA, Tarpy DA, Wilkes JT, vanEngelsdorp D (2014) A national survey of managed honey bee 2012–2013 annual colony losses in the USA: results from the bee informed partnership. J Apic Res 53(1):1–18. https://doi.org/10.3896/ibra.1.53.1.01

Stipp D (2007) As bees go missing a $9.3B crisis lurks. Cable News Network. Available http://money.cnn.com/magazines/fortune/fortune_archive/2007/09/03/100202647/index.htm. Accessed 11 Sept 2018

vanEngelsdorp D, Underwood R, Caron D, Hayes J (2007) An estimate of managed colony losses in the winter of 2006–07: a report commissioned by the Apiary Inspectors of America. Am Bee J 147(7):599–603

vanEngelsdorp D, Hayes J, Underwood RM, Pettis J (2008) A survey of honey bee colony losses in the U.S., fall 2007 to spring 2008. PLoS ONE 3(12):e4071. https://doi.org/10.1371/journal.pone.0004071

vanEngelsdorp D, Evans JD, Saegerman C, Mullin C, Haubruge E, Nguyen BK, Frazier M, Frazier J, Cox-Foster D, Chen Y, Underwood R, Tarpy DR, Pettis JS (2009) Colony collapse disorder: a descriptive study. PLoS ONE 4(8):e6481. https://doi.org/10.1371/journal.pone.0006481

vanEngelsdorp D, Hayes J, Underwood RM, Pettis JS (2010) A survey of honey bee colony losses in the United States, fall 2008 to spring 2009. J Apic Res 49(1):7–14. https://doi.org/10.3896/ibra.1.49.1.03

Williams GR, Tarpy DR, vanEngelsdorp D, Chauzat MP, Cox-Foster DL, Delaplane KS, Neumann P, Pettis JS, Rogers REL, Shutler D (2010) Colony collapse disorder in context. BioEssays 32(10):845–846. https://doi.org/10.1002/bies.201000075

Chapter 4
Food Adulteration Episodes. The Impact of Frauds in the American Market of Dairy Raw Materials

Abstract One of the most important challenges in the current ambit of food safety is the prevention of food frauds. Food fraud is an intentional adulteration that is considered as one of the components of food defence. National and international legislations have progressively put in place preventive action against intentional adulteration through the definition of counterfeiting as a criminal act. The scope of food fraud includes mislabelling, incorrect composition of foods and beverages, non-allowed processes or treatments, and incorrect documentation. In relation to the USA, the recent Food Safety Modernization Act has explicitly considered the problem of intentional adulteration at various levels. With concern to milk, dairy, and cheese products, five different fraud categories are discussed in this chapter. However, future issues may include updates on the FDA proposed revised labelling process when speaking of the use of the term milk for non-milk products such as soy milk or almond milk. This and other matters are currently in progress.

Keywords Corrective action · Documentation · Economically motivated adulteration · Melamine · Mislabelling · Mitigation strategy · Verification

Abbreviations

EMA	Economically motivated adulteration
EU	European Union
FD&C	Federal Food, Drug, and Cosmetic Act
FDA	Food and Drug Administration
FSMA	Food Safety Modernization Act
IA	Intentional adulteration
RASFF	Rapid Alert System for Food and Feed
TRACES	TRAde Control and Expert System
UK	United Kingdom
USA	United States of America
USP	United States Pharmacopeia

© The Author(s), under exclusive license to Springer Nature Switzerland AG 2019   43
S. D. Sharma et al., *Raw Material Scarcity and Overproduction in the Food Industry*,
Chemistry of Foods, https://doi.org/10.1007/978-3-030-14651-1_4

4.1   Food Frauds Today. Prevention of Intentional Adulteration

One of the most important challenges in the current ambit of food safety is the prevention of 'food frauds'. These words concern all possible types of deception using edible materials for human consumptions, with the aim of obtaining economic gains (Everstine et al. 2013; Moore et al. 2012; MSU 2018; Spink and Moyer 2017; Zhang and Xue 2016).

Food fraud is considered as one of the areas in the complex ambit of food safety (Manning 2016). In detail, food protection may be considered as the union of three different strategies with the objective of protecting consumers (Fig. 4.1):

(1)   Food safety—the protection against chemical, microbiological, physical, radiological, or other health hazards directly or indirectly derived from production, packaging, stabilisation, and delivery of foods and beverages.
(2)   Food defence—the protection against malicious actions such as terrorism, bioterrorism, and 'agroterrorism' (Bénoliel 2007; British Standards Institution 2017; Food and Drug Administration 2013; Johnson 2014; Mitenius et al. 2014; United States Congress 2002; World Health Organization 2008).
(3)   Food fraud prevention—the preventive protection against acts of economically motivated adulteration (EMA) in the market of foods and beverages, with impor-

Fig. 4.1 Food fraud is considered as one of the areas in the complex ambit of food protection strategies, with food safety—the protection against health menaces directly or indirectly derived from production, manipulation, packaging, stabilisation, and delivery of foods and beverages, and food defence—the protection against malicious actions such as deliberate terrorism, bioterrorism, and 'agroterrorism' acts against the public health. On the other hand, food fraud prevention is the preventive protection against acts of EMA episodes in the market of foods and beverages. The 'food quality' is not mentioned in this context

tant consequences on the food supply (Applebaum 2014; Sharma and Paradakar 2010).

'Food quality' is not mentioned in this context, although it should be considered part of the problem (Johnson 2014). However, the unintentional characterisation of this area and the substantial economic implications may suggest the analysis of these problems without relation to the regulatory ambit, as seen in the USA when speaking of actions related to factors requiring a preventive control (Ostroff 2017).

The role of national legislations is critical in this ambit because each preventive action against EMA episodes needs the help of police forces (Bogadi et al. 2016; European Union Agency for Law Enforcement 2015). The law enforcement agency in the European Union (EU), Europol, defines counterfeiting as 'an infringement related to industrial property violation' (Montes Saavedra 2014). This example may be useful when considering EMA as a criminal act.

The European Union has considered the problem since 2013, the year of horse meat scandal in the UK (Council of the European Union 2013; European Commission 2013; European Parliament 2013; General Secretariat of the European Council 2014). As a result, the EU action has progressively discovered different frauds concerning foods and beverages by means of systems such as 'TRAde Control and Expert System' (TRACES) and Rapid Alert System for Food and Feed (RASFF) (Parisi et al. 2016). In relation to the last years (2016 and 2017), the EU has reported the following causes related to 792 total food fraud episodes (European Commission 2017):

(a) Mislabelling: 392
(b) Different composition of the food because of: dilution, or replacement, or addition, or removal, of ingredients or chemical components: 182
(c) Non-allowed process(es) or treatment(s): 130
(d) Nonconformities concerning documentation: 88.

As a result, the percentage composition seems to be in favour of mislabelling nonconformities (49.5%), while the remaining categories [(b), (c), and (d)] account for 23.0, 16.4, and 11.1%, respectively (Fig. 4.2). This is related to the action taken by the EU because the reported data does not take into account the activity of Member States at the national level only.

In relation to the USA, the recent Food Safety Modernization Act (FSMA) has explicitly considered the problem of intentional adulteration at various levels, including EMA episodes and food defence-related actions (Bhagat et al. 2016; Kheradia and Warriner 2013; Johnson 2014; Marriott et al. 2018; Spink et al. 2017; Strauss 2011). Although EMA has been considered an act of intentional adulteration, for the context of the intentional adulteration (IA) guidance document, EMA is intended to obtain an economic gain and is not considered to be a risk in the spectrum of IA. By the legal angle, the 'FSMA Final Rule for Mitigation Strategies to Protect Food Against Intentional Adulteration' (US Food and Drug Administration 2018) applies to both domestic and non-US companies obligated to be registered with the Food and Drug Administration (FDA) as food facilities, in accordance with the Federal Food, Drug,

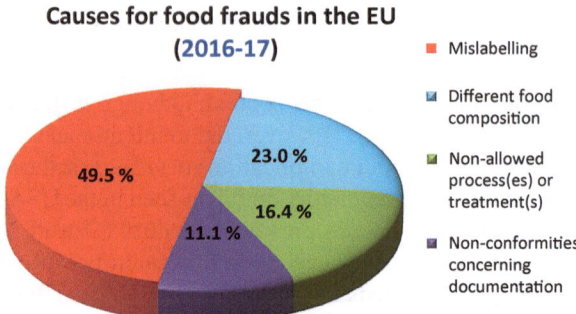

Fig. 4.2 European Union has considered the adulteration problem since 2013, the year of horse meat scandal in the UK, and progressively discovered different frauds concerning foods and beverages by means of systems such as TRACES and RASFF. In relation to the global characterisation of frauds in the EU, a percentage representation of cases in the 2016–17 period is displayed in this Figure depending on the peculiar cause. Shown data—adapted from 'The EU Food Fraud Network and the System for Administrative Assistance & Food Fraud. Annual Report 2017' (European Commission 2017)—concern only the action directly taken by the EU

and Cosmetic (FD&C) Act. However, IA rule does not cover any specific requirements for farms that produce milk (Section 420 of the FD&C Act). Also exempted are the farms that are covered by the standards for produce safety (Section 419 of the FD&C Act) and the farm mixed-type facilities engaged in low-risk activities.

In this ambit, for the IA rule a vulnerability assessment is required, and adequate mitigation strategies (monitoring, corrective actions, and verification) have to be put in place in each processing step—where needed—with the aim of assuring that detected processing vulnerabilities are minimised or prevented (US Food and Drug Administration 2018).

Basically, the Food and Drug Administration (FDA) perception of EMA episodes includes all above-mentioned cases seen in the EU; it should be noted that the related regulatory norms concern health hazards above all because a preventive control is needed (Ostroff 2017).

By the regulatory angle, the FD&C Act mentions 'adulterated foods'—Section 402, (a) and (b)—and these definitions may be expressed as follows, respectively:

(1) Adulterated foods contain one or more substances which may turn the food into a health hazard.
(2) Adulterated foods can also be obtained by means of omission, replacement, addition, or reduction, of one or more of constituents.

On the other hand, 'misbranding' is discussed in Section 403(b) of the FD&C Act (a), (b), and (i), and these definitions may be expressed as follows, respectively:

(1) The mislabelled food is offered for sale.
(2) The food is offered for sale with the name of another food product.
(3) The label fails to bear the common name of the food.

Each of these situations can explain very well why adulterations may be economically motivated: the expected gain is perceived higher if compared with the expected damage when discovered and penalised (Ostroff 2017). The increasing amount of imported foods and beverages (the so-called globalisation) has certainly worsened the problem.

4.2   Intentional Adulteration. Basic Elements

Four main food fraud factors to be considered include (Bouzembrak et al. 2018; Guyader et al. 2018; Josić et al. 2017; Manning 2016; Pallone et al. 2018; Spink et al. 2016):

(1) Characterisation of foods and beverages, with consequent clear identification of the claimed fraud
(2) Identification of drivers related to adulteration episodes
(3) Establishment of an adequate traceability system
(4) Creation, implementation, and amelioration of reliable mechanisms against EMA.

The reliable definition of foods and beverages, with consequent clear identification of the claimed fraud, is needed because of the possible use of the same name for different products. This need is evident when speaking of (Bergeaud-Blackler 2004; Campbell et al. 2011; Ellis et al. 2015; Grundy et al. 2012; Martino et al. 2017):

(a) Similar products obtained with different processes (one of these processing aids is not declared), and/or
(b) Similar foods and beverages produced with different ingredients (one or more of these raw materials is not declared), and/or
(c) Similar products offered for sale and labelled with misleading claims (nutritional data, quality certifications, declared origin to evade taxes, protected brands, declared ethnic and/or religious practices such as *Halal* and *Kosher* accreditations, eco-sustainability or eco-friendly production, etc.).

The identification of drivers related to adulteration episodes is also important: the behaviour of consumers—the real active subject of the purchasing act must be investigated thoroughly, while the viewpoint of adulterating players has been discussed in terms of balance between expected advantage and expected risk (Moyer et al. 2017). On one side, the following features—gender, type of education, and ability to validate/judge food labels by the food consumers—are reported to be critical drivers when speaking of food purchase and related choices. On the other hand, it seems that other important factors indirectly related to key drivers, such as age, professional duties, and purchasing frequency, are not important enough when speaking of undetected EMA by consumers (Charlebois et al. 2016).

The creation, implementation, and continuous improvement of traceability systems are necessary, when speaking of regulatory requisites, both in the EU and the

USA. Some research and practical applications have demonstrated that the traceability of raw materials, including packaging materials, is possible, provided that an efficient system is created and implemented. The situation in the production of cheeses—normal cheeses, processed or melted cheeses, imitation cheeses and similar description, etc.—may show several difficulties depending on peculiar factors: name of the product, number of processing steps, number, and identification of external suppliers, etc. The possibility of a traceability chain leader can improve the traceability system, although similar systems are not always visible (Mania et al. 2016, 2017, 2018).

Finally, the creation, implementation, and amelioration of reliable mechanisms against EMA should be discussed at least in terms of:

(a) Good production and management procedures
(b) Adequate quality management methods
(c) Reliable corrective actions against vulnerabilities
(d) Explicit responsibility for each player of the food chain and into the single production or manufacturing plant (when speaking of food packaging materials because these containers hold labelling information).

This discussion concerns mainly the sector of quality systems and related certifications (Chen et al. 2014; Manning and Soon 2016; Verhoeff and van Duijn, 2013).

4.3 Intentional Adulteration in the Market of Dairy Raw Materials. A Global Analysis

The sector of milk and dairy products, including cheeses, is one of the most exposed ambits when speaking of EMA episodes, at the global level. EMA can be observed in many foods and beverages. EMA can be observed in many food- and beverage-related ambits. According to the United States Pharmacopeia (USP) Food Fraud Database, the following products can be correlated with EMA episodes between 1998 and 2010 (Moore et al. 2012):

- Oils
- Functional food ingredients
- Natural flavouring complexes
- Meats
- Dairy products,
- Protein-based ingredients
- Milk
- Spices
- Colourants
- Wines and other alcoholic beverages
- Other products
- Chemical flavourants

- Cereal and pulses
- Fish and seafood products
- Fruit juices, concentrates, etc.
- Other food and beverage products.

Interestingly, the amount of estimated EMA episodes ascribed to milk and dairy products reached 14 and 4%, respectively, based on 1054 recorded USP scholar references. Subsequent researches have not significantly modified the incidences of milk and dairy-related EMA as evident from data until 2012 (Johnson 2014). For this reason, the situation of raw materials in the milk and dairy sector has been investigated in recent years, in exclusive relation to the US market.

4.4 Intentional Adulteration in the American Market of Dairy Raw Materials

The most known case of EMA related to milk and dairy products is probably the 2008—'melamine' scandal in China (Chan et al. 2008; Ghazi-Tehrani and Pontell 2015; Pei et al. 2011; Schoder 2010). In detail, the arrival of melamine-contaminated high-protein feed and milk-based products in the USA and in other countries from China caused a notable number of sick babies, hospitalisations, and infant death episodes by the late 2008 in China (Yang et al. 2009). In USA, a huge number of cats and dogs were affected by renal failure attributed to melamine presence in pet foods. The diffusion of contaminated foods with melamine through contaminated milk powder has been also demonstrated in other countries such as Tanzania as the result of both legal and illegal business by means of normal and informal channels, where it affected many children (Schoder 2010). This EMA episode has clearly explained the role of globalised markets, although EMA is nothing new when speaking of milk and dairy adulterations.

In general, the most known EMA episodes associated with milk, dairy, and cheese products may be summarised as follows (Fig. 4.3):

(a) Increase in milk protein content in the raw materials (milk) or intermediates for cheese or yogurt products
(b) Partial addition of different and undeclared milk
(c) Modification of milk, dairy, and cheese
(d) Reuse of expired milk, dairy, or cheese products in the production of new milk, dairy, or cheese products (including also processed cheeses and cheese imitations, explicitly produced with 'old' raw materials)
(e) False declaration of origin (geographic production, etc.)
(f) Increase in non-protein amount in certain products, e.g. dilution of milk powders with maltodextrin; replacement or augment of milk lipids with vegetable oils in milks, cheeses, etc. (Johnson 2014).

Fig. 4.3 Most known EMA episodes associated with milk, dairy, and cheese products. The increase in milk proteins can be easily obtained by means of the addition of reconstituted milk powder, rennet, urea, and skim milk powder. The use of different milks from sheep, goat, or buffalo milk has been demonstrated and reported when speaking of addition to cow milk. The same thing should be affirmed when speaking of declared origin and other claims. The modification of shelf life in this sector is particularly important. In addition, the possible reuse or recycling of these cheeses or different products in the production of new milk, dairy, or analogue cheeses should be considered

The increase in milk proteins can be easily obtained by means of the addition of reconstituted milk powder, rennet, urea, and skim milk powder. The result is the increase in the detectable amount of nitrogen-based molecules; in this ambit, the addition of melamine has to be considered. In addition, this fraudulent activity allows hiding the possible water dilution.

The use of different milks (false declarations) from sheep, goat, or buffalo milk has been demonstrated and reported when speaking of addition to cow milk. EMA can be ascribed both to cow milk adulterated with other milk types, and to non-cow-milk containing cow milk. The important thing is that the addition of 'foreign' milk is not declared on labels (Johnson 2014). The same thing should be affirmed when speaking of declared origin (the contrast between real Italian *Parmigiano Reggiano* cheese and non-Italian Parmesan cheese is well known), claimed quality certifications, declaration of peculiar ethnic or religious practices, etc.

The modification of shelf life in this sector is particularly important when speaking of infant baby formulations, soft and semi-hard cheeses, and other products with high perishability. On the other hand, the problem may potentially concern all possible product types depending on the rapidity of consumption in certain geographical areas or commercial circuits; the slower the selling capability for a peculiar product, the higher the probability of shelf-life modification (Johnson 2014). The possible reuse or recycling of these cheeses or different products (including long-durability by-products such as rennet caseins, caseinates, and so on) in the production of new milk, dairy, or analogue cheeses should be considered (Mania et al. 2018; Parisi 2006).

It has to be noted that expired products can easily show microbial contamination, similar to reported cases concerning adulterated milk and milk products because of *Salmonella* and *Listeria* presence (Food and Drug Administration 2016).

Probably, the main menaces to the market of raw materials in the USA are mentioned in the above-shown list. The basic reason is that the global market has imposed the continuous flow of milk and milk-related raw materials worldwide, and the USA is not exposed to this flow differently than other European or non-EU Countries. Consequently, the risk of EMA episodes in the USA is easily correlated with one or more of the above-discussed risks.

Another current issue is the use of the term 'milk' for non-milk products such as soy milk or almond milk, but this matter is currently in progress. At present, the FDA is reported to plan some action plans in the next future (Anonymous 2018). In addition, it has been asked by the International Dairy Foods Association (IDFA) to extend compliance dates for the correct application of 'Nutrition Facts label and Serving Size' final rule until 1 July 2020, for bigger producers, and until 2012 for other producers. This problem concerns the correct labelling process. Other requests have included:

(a)   The amendment of vitamin D fortification levels for milk and milk-based products
(b)   The allowed use of milk protein concentrates and ultra-filtered milk, in accordance with international Codex standards
(c)   The allowed use of ultra-filtered or fluid and dried micro-filtered milk in cheese.

These and other requests to the most recent FDA's request for comments concerning milk regulations are related to the correct labelling of milk, dairy, and cheese products. Substantially, these requests highlight several justified concerns by the milk and dairy producers because of the possible discrepancy between US Regulations and foreign regulations, especially in relation to imported raw materials (International Dairy Foods Association 2018).

References

Anonymous (2018) US Food and Drug Administration question standards of identity on non-dairy products. Food Business Africa, 23 July 2018. Available https://www.foodbusinessafrica.com/2018/07/23/us-food-and-drug-administration-question-standards-of-identity-on-non-dairy-products/. Accessed 12th Sept 2018

Applebaum RS (2014) Terrorism and the nation's food supply perspectives of the food industry: where we are, what we have, and what we need. J Food Sci 69(2):crh48–crh50. https://doi.org/10.1111/j.1365-2621.2004.tb15493.x

Bénoliel I (2007) EU defending food chain against bio-attack. Eur Aff 8(1):70–74

Bergeaud-Blackler F (2004). Social definitions of halal quality: the case of maghrebi muslims in France. In: Harvey M, McMeekin A, Warde A (eds) Qualities of food. Manchester University Press, Manchester and New York

Bhagat A, Caruso G, Micali M, Parisi S (2016) Foods of non-animal origin. Springer International Publishing, Cham. https://doi.org/10.1007/978-3-319-25649-8

Bogadi NP, Banović M, Babić I (2016) Food defence system in food industry: perspective of the EU countries. J Verbrauch Lebensm 11(3):217–226. https://doi.org/10.1007/s00003-016-1022-8

Bouzembrak Y, Steen B, Neslo R, Linge J, Mojtahed V, Marvin HJP (2018) Development of food fraud media monitoring system based on text mining. Food Control 93:283–296. https://doi.org/10.1016/j.foodcont.2018.06.003

British Standards Institution (2017) PAS 96:2014 guide to protecting and defending food and drink deliberate attack, 4th edn, Nov 2017. The British Standards Institution, London. Available http://www.food.gov.uk/sites/default/files/pas96-2014-food-drink-protection-guide.pdf. Accessed 12th Sept 2018

Campbell H, Murcott A, MacKenzie A (2011) Kosher in New York City, Halal in Aquitaine: challenging the relationship between neoliberalism and food auditing. Agric Hum Val 28:67–79

Chan EYY, Griffiths SM, Chan CW (2008) Public-health risks of melamine in milk products. Lancet 372(9648):1444–1445. https://doi.org/10.1016/S0140-6736(08)61604-9

Charlebois S, Schwab A, Henn R, Huck CW (2016) Food fraud: an exploratory study for measuring consumer perception towards mislabeled food products and influence on self-authentication intentions. Trends Food Sci Technol 50:211–218. https://doi.org/10.1016/j.tifs.2016.02.003

Chen C, Zhang J, Delaurentis T (2014) Quality control in food supply chain management: an analytical model and case study of the adulterated milk incident in China. Int J Prod Econ 152:188–199. https://doi.org/10.1016/j.ijpe.2013.12.016

Council of the European Union (2013) Council conclusions on setting the EU's priorities for the fight against serious and organised crime between 2014 and 2017. Justice and Home Affairs Council meeting—Luxembourg, 6 and 7 June 2013, 5 p. Council of the European Union, Brussels. Available http://www.consilium.europa.eu/uedocs/cms_data/docs/pressdata/en/jha/137401.pdf. Accessed 12th Sept 2018

European Commission (2013) Commission Recommendation of 19 February 2013 on a coordinated control plan with a view to establish the prevalence of fraudulent practices in the marketing of certain foods. Off J Eur Union L48:28–32

Department of Food Safety, Zoonoses and Foodborne Disease, Cluster on Health Security and Environment, Geneva. Available http://seafood.oregonstate.edu/.pdf%20Links/WHO%20Food%20Safety%20Issues%20-%20Terrorist%20Threats%20to%20Food.pdf. Accessed 12th Sept 2018

Ellis DI, Muhamadali H, Haughey SA, Elliott CT, Goodacre R (2015) Point-and-shoot: rapid quantitative detection methods for on-site food fraud analysis–moving out of the laboratory and into the food supply chain. An Methods 7(22):9401–9414. https://doi.org/10.1039/c5ay02048d

European Commission (2017) The EU Food fraud network and the system for administrative assistance & food fraud. Annual Report 2017. European Commission, Brussels. Available https://ec.europa.eu/food/sites/food/files/safety/docs/food-fraud_network_activity_report_2017.pdf. Accessed 12th Sept 2018

European Parliament (2013) Report on the food crisis, fraud in the food chain and the control thereof (2013/2091(INI). A7-0434/2013, 4.12.2013. European Parliament, Brussels. Available http://www.europarl.europa.eu/sides/getDoc.do?pubRef=-//EP//TEXT+REPORT+A7-2013-0434+0+DOC+XML+V0//EN. Accessed 12th Sept 2018

European Union Agency for Law Enforcement (2015) Record seizures of fake food and drink in Interpol-Europol operation. Press Release, 16 February 2015. European Union Agency for Law Enforcement (Europol), The Hague. Available https://www.europol.europa.eu/content/record-seizures-fake-food-and-drink-interpol-europol-operation. Accessed 12th Sept 2018

Everstine K, Spink J, Kennedy S (2013) Economically motivated adulteration (EMA) of food: common characteristics of EMA incidents. J Food Prot 76(4):723–735. https://doi.org/10.4315/0362-028x.jfp-12-399

Food and Drug Administration (2013) Focused mitigation strategies to protect food against intentional adulteration; proposed rule. Fed Regist 78(247):78013–78061 (FR Doc No: 2013-30373). Available https://www.gpo.gov/fdsys/pkg/FR-2013-12-24/html/2013-30373.htm. Accessed 12th Sept 2018

Food and Drug Administration (2016) Food regulators seize adulterated milk products for food safety violations. News Release, 30 Nov 2016. Food and Drug Administration, Washington, DC. Available https://www.fda.gov/newsevents/newsroom/pressannouncements/ucm531188.htm. Accessed 12th Sept 2018

General Secretariat of the European Council (2014) Draft council conclusions on the role of law enforcement cooperation in combating food crime. Brussels, 27 November 2014. Council of the European Union, Brussels. Available http://data.consilium.europa.eu/doc/document/ST-15623-2014-INIT/en/pdf. Accessed 12th Sept 2018

Ghazi-Tehrani AK, Pontell HN (2015) Corporate crime and state legitimacy: the 2008 Chinese melamine milk scandal. Crime Law Soc Ch 63(5):247–267. https://doi.org/10.1007/s10611-015-9567-5

Grundy HH, Kelly SD, Charlton AJ, Donarski JA, Hird SJ, Collins MJ (2012) Food authenticity and food fraud research: achievements and emerging issues. J Assoc Pub An 40:65–68. Available http://www.apajournal.org.uk/2012_0065-0068.pdf. Accessed 12th Sept 2018

Guyader S, Thomas F, Portaluri V, Jamin E, Akoka S, Silvestre V, Remaud G (2018) Authentication of edible fats and oils by non-targeted 13 C INEPT NMR spectroscopy. Food Control 91:216–224. https://doi.org/10.1016/j.foodcont.2018.03.046

International Dairy Foods Association (2018) IDFA recommends areas for regulatory reform for dairy product manufacturing. International Dairy Foods Association (IDFA), Washington, DC, 05 Feb 2018. Available https://www.idfa.org/news-views/news-releases/article/2018/02/05/idfa-recommends-areas-for-regulatory-reform-for-dairy-product-manufacturing. Accessed 12th Sept 2018

Johnson R (2014) Food fraud and "economically motivated adulteration" of food and food ingredients. Congressional Research Service, Washington, DC. Available https://fas.org/sgp/crs/misc/R43358.pdf. Accessed 12th Sept 2018

Josić D, Peršurić Ž, Rešetar D, Martinović T, Saftić L, Kraljević Pavelić S (2017) Use of foodomics for control of food processing and assessing of food safety. Adv Food Nutr Res 81:187–229. https://doi.org/10.1016/bs.afnr.2016.12.001

Kheradia A, Warriner K (2013) Understanding the food safety modernization act and the role of quality practitioners in the management of food safety and quality systems. TQM J 25(4):347–370. https://doi.org/10.1108/17542731311314854

Mania I, Barone C, Caruso G, Delgado A, Micali M, Parisi S (2016) Traceability in the cheesemaking field. The Regulatory ambit and practical solutions. Food Qual Mag 3:18. Available https://www.joomag.com/magazine/food-quality-magazine-july-2016/0123592001469962176. Accessed 12th Sept 2018

Mania I, Barone C, Pellerito A, Laganà P, Parisi S (2017) Trasparenza e Valorizzazione delle Produzioni Alimentari. L'etichettatura e la Tracciabilità di Filiera come Strumenti di Tutela delle Produzioni Alimentari. Ind Aliment 56(581):18–22

Mania I, Delgado AM, Barone C, Parisi S (2018) Traceability in the dairy industry in Europe. Springer International Publishing, Cham, in press. https://doi.org/10.1007/978-3-030-00446-0

Manning L (2016) Food fraud: policy and food chain. Curr Opinion Food Sci 10:16–21. https://doi.org/10.1016/j.cofs.2016.07.001

Manning L, Soon JM (2016) Food safety, food fraud, and food defense: a fast evolving literature. J Food Sci 81(4):R823–R834. https://doi.org/10.1111/1750-3841.13256

Marriott NG, Schilling MW, Gravani RB (2018) Principles of food sanitation. Springer International Publishing, Cham. https://doi.org/10.1007/978-3-319-67166-6

Martino G, Karantininis K, Pascucci S, Dries L, Codron JM (eds) (2017) It's a jungle out there—the strange animals of economic organization in agri-food value chains. Wageningen Academic Publishers, Wageningen. https://doi.org/10.3920/978-90-8686-844-5

Mitenius N, Kennedy SP, Busta FF (2014) Food defense. In: Motarjemi Y, Lelieveld H (eds) Food safety management—a practical guide for the food industry. Academic Press, San Diego

Montes Saavedra E (2014) From 'food to fraud. The continuous battle against dishonest practices in the food chain. A comparative analysis between the European and the American (USA) food fraud control systems. Dissertation, Wageningen University, Wageningen. Available http://edepot.wur.nl/304294. Accessed 12th Sept 2018

54   4 Food Adulteration Episodes. The Impact of Frauds …

Moore JC, Spink J, Lipp M (2012) Development and application of a database of food ingredient fraud and economically motivated adulteration from 1980 to 2010. J Food Sci 77(4):R118–R126. https://doi.org/10.1111/j.1750-3841.2012.02657.x

Moyer DC, DeVries JW, Spink J (2017) The economics of a food fraud incident–case studies and examples including Melamine in Wheat Gluten. Food Control 71:358–364. https://doi.org/10.1016/j.foodcont.2016.07.015

MSU (2018) Food fraud reference sheet. Michigan State University (MSU), College of Veterinary Medicine, East Lansing. Available http://foodfraud.msu.edu/food-fraud-reference-sheet/. Accessed 12th Sept 2018

Ostroff S (2017) A regulator's view on preventing and mitigating economic adulteration of food. In: Food fraud conference, Quebec City, 5 Apr 2017

Pallone JAL, Caramês ETS, Alamar PD (2018) Green analytical chemistry applied in food analysis: alternative techniques. Curr Opinion Food Sci 22:115–121. https://doi.org/10.1016/j.cofs.2018.01.009

Parisi S (2006) Profili chimici delle caseine presamiche alimentari. Ind Aliment 45(457):377–383

Parisi S, Barone C, Sharma RK (2016) Chemistry and food safety in the EU. Springer International Publishing, Heidelberg, Germany. https://doi.org/10.1007/978-3-319-33393-9

Pei X, Tandon A, Alldrick A, Giorgi L, Huang W, Yang R (2011) The China melamine milk scandal and its implications for food safety regulation. Food Policy 36(3):412–420. https://doi.org/10.1016/j.foodpol.2011.03.008

Schoder D (2010) Melamine milk powder and infant formula sold in East Africa. J Food Prot 73(9):1709–1714. https://doi.org/10.4315/0362-028x-73.9.1709

Sharma K, Paradakar M (2010) The melamine adulteration scandal. Food Sec 2(1):97–107. https://doi.org/10.1007/s12571-010-0057-4

Spink J, Moyer DC (2017) Food fraud prevention—how to start how much is enough? New Food 20(1):13–16

Spink J, Moyer DC, Whelan P (2016) The role of the public private partnership in Food Fraud prevention—includes implementing the strategy. Curr Opinion Food Sci 10:68–75. https://doi.org/10.1016/j.cofs.2016.10.002

Spink J, Ortega DL, Chen C, Wu F (2017) Food fraud prevention shifts the food risk focus to vulnerability. Trends Food Sci Technol 62:215–220. https://doi.org/10.1016/j.tifs.2017.02.012

Strauss DM (2011) An analysis of the FDA food safety modernization act: protection for consumers and boon for business. Food Drug Law J 66(3):353–374

United States Congress (2002) Public health security and bioterrorism preparedness and response act of 2002. Public Law 107–188, H.R. 3448, June 2002. United States Congress, Washington, DC. Available https://www.gpo.gov/fdsys/pkg/PLAW-107publ188/pdf/PLAW-107publ188.pdf. Accessed 12th Sept 2018

U.S. Food and Drug Administration (2018) FSMA final rule for mitigation strategies to protect food against intentional adulteration. U.S. Food and Drug Administration, Washington, DC. Available https://www.fda.gov/food/guidanceregulation/fsma/ucm378628.htm?source=go#coverage. Accessed 12th Sept 2018

Verhoeff M, van Duijn G (2013) Quality and food safety assurance and control. In: Hamm W, Hamilton RJ, Calliauw G (eds) Edible oil processing. Wiley & Sons Ltd, New York. https://doi.org/10.1002/9781118535202.ch9

World Health Organization (2008) Terrorist threats to food: guidance for establishing and strengthening prevention and response systems, revision May 2008. World Health Organization, Geneva

Yang R, Huang W, Zhang L, Thomas M, Pei X (2009) Milk adulteration with melamine in China: crisis and response. Qual Ass Saf Crops Food 1(2):111–116. https://doi.org/10.1111/j.1757-837X.2009.00018.x

Zhang W, Xue J (2016) Economically motivated food fraud and adulteration in China: an analysis based on 1553 media reports. Food Control 67:192–198. https://doi.org/10.1016/j.foodcont.2016.03.004